高职高专教育"十二五"规划教材

# 交换机/路由器配置与管理实验教程

主　编　伍技祥　张　庚

副主编　韩桂萍　向　涛　单科峰

中国水利水电出版社
www.waterpub.com.cn

## 内容提要

本书根据网络工程实际工作过程所需要的知识和技能抽象出 21 个教学实验。内容上分两个部分：第一部分（实验一到实验十三）为交换机实验，包含交换机的几种模式的认识、交换机的口令和登录的配置、交换机的端口配置、VLAN 的配置、生成树协议的配置、VTP 的配置、链路聚合的配置、静态路由的配置、NAT 的配置、DHCP 的配置、VRRP 的配置、HSRP 的配置、交换机综合实验；第二部分（实验十四到实验二十一）为路由器实验，包含路由器 SSH 配置、访问列表、RIP 路由协议的配置、OSPF 路由协议的配置、策略路由的配置、PPP 协议的配置、帧中继的配置、路由器配置综合实验。附录为"Dynamips 模拟器的使用"，供没有使用过 Dynamips 模拟器的用户参考。

本书以知识"必需、够用"为原则，从职业岗位分析入手，展开实验教学内容，强化学生的技能训练，在训练过程中巩固所学知识。本书既可以作为高职院校网络技术实训教材，也可以作为社会培训教材。

本书配有免费电子教案和实验素材，读者可以从中国水利水电出版社网站以及万水书苑下载，网址为：http://www.waterpub.com.cn/softdown/或 http://www.wsbookshow.com。

## 图书在版编目（CIP）数据

交换机/路由器配置与管理实验教程 / 伍技祥，张庚主编. -- 北京：中国水利水电出版社，2013.1
高职高专教育"十二五"规划教材
ISBN 978-7-5170-0296-3

Ⅰ. ①交… Ⅱ. ①伍… ②张… Ⅲ. ①计算机网络－信息交换机－高等职业教育－教材②计算机网络－路由选择－高等职业教育－教材 Ⅳ. ①TN915.05

中国版本图书馆CIP数据核字(2012)第253656号

策划编辑：雷顺加　　责任编辑：张玉玲　　加工编辑：孙 丹　　封面设计：李 佳

| 书　名 | 高职高专教育"十二五"规划教材<br>交换机/路由器配置与管理实验教程 |
|---|---|
| 作　者 | 主　编　伍技祥　张　庚<br>副主编　韩桂萍　向　涛　单科峰 |
| 出版发行 | 中国水利水电出版社<br>（北京市海淀区玉渊潭南路1号D座　100038）<br>网址：www.waterpub.com.cn<br>E-mail: mchannel@263.net（万水）<br>　　　　sales@waterpub.com.cn<br>电话：（010）68367658（发行部）、82562819（万水） |
| 经　售 | 北京科水图书销售中心（零售）<br>电话：（010）88383994、63202643、68545874<br>全国各地新华书店和相关出版物销售网点 |
| 排　版 | 北京万水电子信息有限公司 |
| 印　刷 | 三河市铭浩彩色印装有限公司 |
| 规　格 | 184mm×260mm　16开本　16印张　403千字 |
| 版　次 | 2013年1月第1版　2013年1月第1次印刷 |
| 印　数 | 0001—3000册 |
| 定　价 | 28.00元 |

凡购买我社图书，如有缺页、倒页、脱页的，本社发行部负责调换

**版权所有·侵权必究**

# 前　　言

随着计算机和网络技术的迅猛发展，计算机网络及应用已经渗透到社会各个领域，各行各业都处在全面网络化和信息化建设进程中，对网络应用型人才的需求也与日俱增，计算机网络行业已成为技术人才稀缺的行业之一。

传统网络专业课程设置不能满足用人单位对网络岗位的技能要求。传统网络专业课程设置受网络实训设备、师资和教材的限制，未能开设交换机和路由器配置与管理方面的课程，或者只是教师进行演示，学生无法动手实际操作；主要侧重于网络理论介绍和针对小型网络组建的技能培养，这就造成了所培养的毕业生只能胜任小型网络的组建和维护管理工作。但目前一般的局域网都需要划分子网，这就要求网络管理者必须掌握对网络核心设备的配置和管理能力。

没有掌握交换机和路由器的配置与管理这一网络核心技能，是造成目前计算机网络专业毕业生就业难的主要原因之一，也是造成目前网络专业就业市场出现"两旺两难"的主要原因。"两旺"是指供方和需求都旺盛；"两难"是指学生就业难，企业招人难。网络工程的组建，除了网络规划设计、网络综合布线以外，最关键的任务就是对交换机和路由器按网络规划设计的要求进行合理配置，这是技术含量较高的一个工作岗位。

本书是一本全面、面向结果的模块化项目式网络实验教材，旨在提供关于计算机网络方面所涉及的主要知识体系和知识内容。全书根据网络工程实际工作过程所需要的知识和技能抽象出了 21 个教学实验。从内容上分两个部分：第一部分（实验一到实验十三）为交换机实验，包含交换机的几种模式的认识、交换机的口令和登录的配置、交换机的端口配置、VLAN 的配置、生成树协议的配置、VTP 的配置、链路聚合的配置、静态路由的配置、NAT 的配置、DHCP 的配置、VRRP 的配置、HSRP 的配置、交换机综合实验；第二部分（实验十四到实验二十一）为路由器实验，包含路由器 SSH 配置、访问列表、RIP 路由协议的配置、OSPF 路由协议的配置、策略路由的配置、PPP 协议的配置、帧中继的配置、路由器配置综合实验（将前面交换机、路由器所有相关的知识进行整合，通过该综合实验使学生真正掌握交换机路由器配置，在实际使用中灵活地应用）。

通过全书提供的理论知识和实验知识模块体系，使读者全面感受网络知识和实际应用的有机结合，使知识和实际应用建立起对应关系，使学习更具针对性。同时本书提供的实验项目可以有效帮助读者理解课堂教学中学到的抽象理论并加深印象。

常见的 Cisco 设备网络模拟器有 Parker Tracer、Boson Netsim、Dynamips。本书选择 Dynamips 模拟器，主要因为 Dynamips 是模拟 Cisco 真实的硬件设备，加载 Cisco 的映像文件能够使用真实设备上的所有命令；能够将模拟环境与真实的网络环境进行互连；能够抓取真实的网络数据包进行分析。读者在读此书之前，如果没有使用过 Dynamips 模拟器，请参见附录"Dynamips 模拟器的使用"。

本书的特点主要体现在以下四个方面：
- 总体采用任务驱动、项目教学方式进行组织编写。

- 突出计算机网络规划设计、网络工程组建和网络管理能力的培养。
- 案例经典,可操作性强。
- 内容上由易到难,以循序渐进的方式组织内容,最后通过综合实例将前面的教学内容进行综合应用。

本书分工如下:由伍技祥编写实验十三、实验十五至实验十九及实验二十一,张庚编写实验一至实验四,韩桂萍编写实验五至实验八,向涛编写实验九至实验十二,单科峰编写实验二十和附录。

### 排版约定

实验需要掌握的命令部分的内容约定:

斜体:为用户定义的参数,必须满足系统要求。

[](方括号):中括号中的内容为可选内容,不能输入方括号。

{}(大括号):大括号中的内容为必选内容,不能输入大括号。

|(竖线):表示分隔符,用于分开可选择的选项。

#(井号):以#开头的为注释行,是对上一行命令的解释。

本书得到中国水利水电出版社万水分社的大力支持和帮助,在此致以衷心的感谢!由于编者水平有限,书中难免有不妥和错误之处,恳请广大读者批评指正。

编 者

2012 年 9 月

# 目　　录

前言

**实验一　交换机的几种模式的认识** ………… 1
　　一、实验目的 ……………………… 1
　　二、实验内容 ……………………… 1
　　三、实验需要掌握的命令 ………… 1
　　四、实验拓扑图和网络文件 ……… 2
　　五、实验步骤 ……………………… 2

**实验二　交换机的口令和登录的配置** ……… 4
　　一、实验目的 ……………………… 4
　　二、实验内容 ……………………… 4
　　三、实验需要掌握的命令 ………… 4
　　四、实验拓扑图和网络文件 ……… 5
　　五、实验配置步骤 ………………… 5
　　六、实验测试和查看相关配置 …… 6
　　七、注意问题 ……………………… 8

**实验三　交换机的端口配置** ………………… 9
　　一、实验目的 ……………………… 9
　　二、实验要内容 …………………… 9
　　三、实验需要掌握的命令 ………… 9
　　四、实验拓扑图和网络文件 ……… 10
　　五、实验配置步骤 ………………… 10
　　六、实验测试和查看相关配置 …… 11
　　七、注意问题 ……………………… 12

**实验四　VLAN 的配置** ……………………… 13
　　一、实验目的 ……………………… 13
　　二、实验内容 ……………………… 13
　　三、实验原理 ……………………… 13
　　四、实验需要掌握的命令 ………… 17
　　五、实验拓扑图和网络文件 ……… 18
　　六、实验配置步骤 ………………… 19
　　七、实验测试和查看相关配置 …… 22
　　八、注意问题 ……………………… 23

**实验五　生成树协议的配置** ………………… 24
　　一、实验目的 ……………………… 24

　　二、实验内容 ……………………… 24
　　三、实验原理 ……………………… 24
　　四、实验需要掌握的命令 ………… 25
　　五、实验拓扑图和网络文件 ……… 26
　　六、实验配置步骤 ………………… 27
　　七、实验测试和查看相关配置 …… 29
　　八、注意问题 ……………………… 34

**实验六　VTP 的配置** ………………………… 35
　　一、实验目的 ……………………… 35
　　二、实验内容 ……………………… 35
　　三、实验原理 ……………………… 35
　　四、实验需要掌握的命令 ………… 36
　　五、实验拓扑图和网络文件 ……… 37
　　六、实验配置步骤 ………………… 39
　　七、实验测试和查看相关配置 …… 41
　　八、注意问题 ……………………… 43

**实验七　链路聚合的配置** …………………… 44
　　一、实验目的 ……………………… 44
　　二、实验内容 ……………………… 44
　　三、工作原理 ……………………… 44
　　四、实验需要掌握的命令 ………… 45
　　五、实验拓扑图与网络文件 ……… 45
　　六、实验配置步骤 ………………… 46
　　七、实验测试和查看相关配置 …… 47
　　八、注意问题 ……………………… 49

**实验八　静态路由的配置** …………………… 50
　　一、实验目的 ……………………… 50
　　二、实验内容 ……………………… 50
　　三、实验原理 ……………………… 50
　　四、实验需要掌握的命令 ………… 51
　　五、实验拓扑图和网络文件 ……… 51
　　六、实验配置步骤 ………………… 53
　　七、实验测试和查看相关配置 …… 56

八、注意问题 …………………………………… 59
实验九　NAT 的配置 …………………………………… 60
　　一、实验目的 …………………………………… 60
　　二、实验内容 …………………………………… 60
　　三、实验原理 …………………………………… 60
　　四、实验需要掌握的命令 …………………… 61
　　五、实验拓扑图和网络文件 ………………… 62
　　六、实验配置步骤 ……………………………… 63
　　七、实验测试和查看相关配置 ……………… 67
　　八、注意问题 …………………………………… 69
实验十　DHCP 的配置 ………………………………… 70
　　一、实验目的 …………………………………… 70
　　二、实验内容 …………………………………… 70
　　三、实验原理 …………………………………… 70
　　四、实验需要掌握的命令 …………………… 71
　　五、实验拓扑图和拓扑文件 ………………… 72
　　六、实验配置步骤 ……………………………… 73
　　七、实验测试和查看相关配置 ……………… 75
　　八、注意问题 …………………………………… 76
实验十一　VRRP 的配置 ……………………………… 77
　　一、实验目的 …………………………………… 77
　　二、实验内容 …………………………………… 77
　　三、VRRP 协议简介 …………………………… 77
　　四、实验需要掌握的命令 …………………… 78
　　五、实验拓扑图和网络文件 ………………… 79
　　六、实验配置步骤 ……………………………… 81
　　七、实验测试和查看相关配置 ……………… 83
　　八、注意问题 …………………………………… 85
实验十二　HSRP 的配置 ……………………………… 86
　　一、实验目的 …………………………………… 86
　　二、实验内容 …………………………………… 86
　　三、实验原理 …………………………………… 86
　　四、实验需要掌握的命令 …………………… 89
　　五、实验拓扑图与网络文件 ………………… 89
　　六、实验配置步骤 ……………………………… 91
　　七、实验测试和查看相关配置 ……………… 94
　　八、注意问题 …………………………………… 96
实验十三　交换机综合实验 …………………………… 97
　　一、实验目的 …………………………………… 97
　　二、实验拓扑图和网络文件 ………………… 97
　　三、IP 地址的规划 …………………………… 101
　　四、实验配置步骤 …………………………… 103
　　五、实验测试和查看相关配置 …………… 114
实验十四　路由器 SSH 配置 ………………………… 117
　　一、实验目的 ………………………………… 117
　　二、实验内容 ………………………………… 117
　　三、实验原理 ………………………………… 117
　　四、实验需要掌握的命令 ………………… 118
　　五、实验拓扑图和拓扑文件 ……………… 119
　　六、实验配置步骤 …………………………… 119
　　七、实验测试和查看相关配置 …………… 120
　　八、注意问题 ………………………………… 121
实验十五　访问列表 …………………………………… 122
　　一、实验目的 ………………………………… 122
　　二、实验内容 ………………………………… 122
　　三、工作原理 ………………………………… 122
　　四、实验需要掌握的命令 ………………… 125
　　五、实验拓扑图和网络文件 ……………… 126
　　六、实验配置步骤 …………………………… 127
　　七、实验测试和查看相关配置 …………… 129
　　八、注意问题 ………………………………… 131
实验十六　RIP 路由协议的配置 …………………… 132
　　一、实验目的 ………………………………… 132
　　二、实验内容 ………………………………… 132
　　三、实验原理 ………………………………… 132
　　四、实验需要掌握的命令 ………………… 136
　　五、实验拓扑图和网络文件 ……………… 138
　　六、实验配置步骤 …………………………… 139
　　七、实验测试和查看相关配置 …………… 144
　　八、注意问题 ………………………………… 149
实验十七　OSPF 路由协议的配置 ………………… 150
　　一、实验目的 ………………………………… 150
　　二、实验内容 ………………………………… 150
　　三、OSPF 协议的简介 ……………………… 150
　　四、实验需要掌握的命令 ………………… 160
　　五、实验拓扑图和网络文件 ……………… 162
　　六、实验配置步骤 …………………………… 164
　　七、实验测试和查看相关配置 …………… 169

八、注意问题 174

**实验十八　策略路由的配置** 175
　　一、实验目的 175
　　二、实验内容 175
　　三、实验原理 175
　　四、实验需要掌握的命令 176
　　五、实验拓扑图和网络文件 176
　　六、实验配置步骤 178
　　七、实验测试和查看相关配置 181
　　八、注意问题 182

**实验十九　PPP 协议的配置** 183
　　一、实验目的 183
　　二、实验内容 183
　　三、实验原理 183
　　四、实验需要掌握的命令 187
　　五、实验拓扑图和拓扑文件 188
　　六、实验配置步骤 188
　　七、实验测试和查看相关配置 190
　　八、注意问题 192

**实验二十　帧中继的配置** 193
　　一、实验目的 193
　　二、实验内容 193
　　三、帧中继工作原理 193
　　四、实验需要掌握的命令 196
　　五、实验拓扑图和网络文件 197
　　六、实验配置步骤 199
　　七、实验测试和查看相关配置 202
　　八、注意问题 203

**实验二十一　路由器配置综合实验** 204
　　一、实验目的 204
　　二、实验内容 204
　　三、实验拓扑图和拓扑文件 204
　　四、IP 地址规划 206
　　五、实验配置步骤 208
　　六、实验测试和查看相关配置 218

**附录　Dynamips 模拟器的使用** 234
**参考文献** 248

# 实验一　交换机的几种模式的认识

## 一、实验目的

掌握交换机的几种模式之间的关系和作用，熟练掌握它们之间的切换。

## 二、实验内容

（1）掌握几种模式之间的关系；
（2）掌握几种模式的作用；
（3）能够熟练地在几种模式之间切换。

## 三、实验需要掌握的命令

switch>enable
！从用户模式进入特权模式
switch#config terminal
！从特权模式进入全局配置模式
switch#vlan database
！从特权模式进入 VLAN 配置模式
switch (config)#interface FastEthernet 1/0
！从特权模式进入端口配置模式
switch (config)#line vty 0 4
！从特权模式进入线路配置模式
switch (config-if)#end
！从端口模式切换到特权模式
switch (config-line)#end
！从线路模式切换到特权模式
switch (config)#end
或者 hostname(config)#exit
！从全局配置模式切换到特权模式
switch (config-if)#exit
！从端口模式切换到全局配置模式
switch (config-line)#exit
！从线路模式切换到全局配置模式
switch (vlan)#exit
！从 VLAN 配置模式切换到特权模式
几种模式之间的切换如图 1-1 所示。

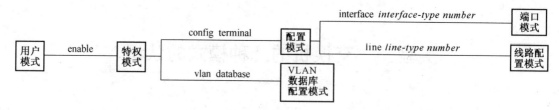

图 1-1　交换机模式切换

在用户模式下输入 enable 命令就会进入特权模式；在特权模式下输入 config terminal 就会进入全局配置模式，输入 vlan database 就会进入 VLAN 数据库配置模式；在全局配置模式下输入 interface *interface-type number* 就会进入端口配置模式，输入 line *line-type number* 就会进入线路配置模式。利用 exit 和 end 命令，可以退回到上一级模式中，其中 VLAN 数据库配置模式和特权模式的退出只能用 exit 命令。

## 四、实验拓扑图和网络文件

（1）网络拓扑图。

网络拓扑图如图 1-2 所示。

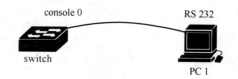

图 1-2　网络拓扑图

说明：在真实的网络环境中，上图中的 console 端口和 RS 232 是配置线将交换机和 PC 主机连接，从而可以通过 PC 对交换机进行配置，在此实验中没有这两个端口。

（2）网络文件。

网络文件 switch1.net 的内容如下：

```
# switch1.net 文件开始
[localhost]       #设置虚拟设备的主机 IP 地址
    [[3640]]      #设置虚拟主机模拟的设备类型
        image = ..\c3640-ik9o3s-mz.123-26.bin      #设置设备加载镜像文件的路径
        ram = 160    #设置设备的内存大小
    [[ROUTER R1]]  #添加设备，类型为路由器，名称为 R1
        model = 3640    #设备的类型为 3640
        slot1 = NM-16ESW   #加载模拟交换机模块
# switch1.net 文件结束
```

## 五、实验步骤

第一步：将鼠标放在 dynamips 安装目录中的 dynamips-start.cmd 文件上，双击鼠标左键，运行 dynamips 安装目录中的 dynamips-start.cmd 文件；此时出现如图 1-3 所示的窗口，不要关闭此窗口，将它最小化；此步的作用为模拟所有交换机的硬件。

第二步：打开 switch1.net 所在的文件夹，将鼠标放在 swith1.net 文件上，双击鼠标左键，运行 switch1.net 网络文件；此时会出现如图 1-4 所示的窗口；此步的作用为向上一步模拟的硬件中加载镜像文件，同时产生模拟设备。

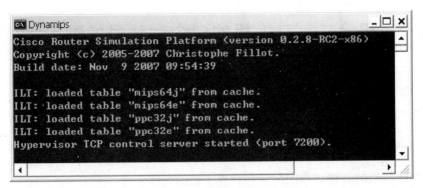

图 1-3　Dynamips 服务器

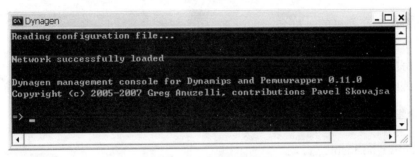

图 1-4　Dynagen 控制台

第三步：在图 1-4 的窗口中输入 console R1 命令，此时出现如图 1-5 所示的窗口。

图 1-5　配置交换机

第四步：接下来就可以在上面的窗口中对交换机进行各种操作了。

# 实验二　交换机的口令和登录的配置

设置交换机的口令是为了增加交换机的安全性，配置远程登录是为了增加交换机管理的方便性。只有同时设置远程登录和 enable 密码（网络互连正常情况下），才能进行远程管理。

## 一、实验目的

熟练掌握交换机的各种口令的配置和设置主机的名称。

## 二、实验内容

（1）熟练配置交换机的主机名；
（2）熟练配置特权模式的口令；
（3）熟练配置控制端口的口令和允许登录；
（4）熟练配置虚拟终端的口令和允许远程登录。

## 三、实验需要掌握的命令

switch(config)#hostname *xx*
！设置交换机的主机名

switch(config)#enable secret *xxx*
！设置特权模式的加密口令

switch(config)#enable password *xxx*
！设置特权模式的非密口令

switch(config)#line console 0
！进入控制台配置模式

switch(config)#line vty 0 4
！进入虚拟终端配置模式

switch(config-line)#login
！允许远程登录

switch(config-line)#password *xx*
！设置登录口令 xx

switch (config-line)# exec-timeout *minutes* [*seconds*]
！定义所有 EXEC 模式会话的空闲时间

switch (config-line)# absolute-timeout *minute*
！定义一条线路的绝对超时时间

switch #show line console 0
！查看 console 端口状态信息

switch #show line vty 0
！查看虚拟终端状态信息

switch #show running-config
！查看交换机的配置信息

switch(config)#interface *interface-type slot/number*
！进入端口

switch(config-if)#ip address *ip-address subnet-mask*
！设置端口 IP 地址

switch(config-if)#no shutdown
！启用当前端口

switch(config-if)#shutdown
！关闭当前端口

switch(config)#write
！将内存中的配置信息写入 flash 中

### 四、实验拓扑图和网络文件

（1）网络拓扑图。

网络拓扑图如图 2-1 所示。

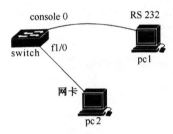

图 2-1 网络拓扑图

（2）网络文件。

网络文件 switch2.net 的内容如下：
```
# switch2.net 文件开始
[localhost]
    [[3640]]
    image = ..\c3640-ik9o3s-mz.123-26.bin
    ram = 160
    [[ROUTER R1]]
    model = 3640
    slot1 = NM-16ESW #加载模拟交换机模块
    f0/0 = NIO_gen_eth:\Device\NPF_{E2B31736-C716-4215-A9CB-ADA78C760F52}
#将上一行 {E2B31736-C716-4215-A9CB-ADA78C760F52} 值修改为你的网卡值，通过运行桌面上的
Network device list.cmd 快捷方式获得
# switch2.net 文件结束
```

### 五、实验配置步骤

（1）设置交换机的主机名。

sw1>enable
sw1#configure terminal

sw1(config)#hostname switch

（2）设置交换机的特权模式的密码。

switch(config)#enable password 123321
switch(config)#enable secret 123456

（3）设置交换机 console 端口的登录密码，同时设置了空闲时间和每一次连接的时间限制。

switch(config)#line console 0
switch(config-line)#password a1a1a1
switch(config-line)#login        #此命令如果没有设置，则无法登录
switch(config-line)#exec-timeout 5
switch(config-line)#absolute-timeout 10
switch(config-line)#exit
switch(config)#exit

（4）设置交换机的虚拟终端登录和登录密码。

switch(config)#line vty 0 4
switch(config-line)#password ababab    #此命令如果没有设置，则虚拟终端无法登录
switch(config-line)#login   #此命令如果没有设置，则 password 命令将失效
switch(config-line)#exec-timeout 5
switch(config-line)#absolute-timeout 10
switch(config-line)#end
switch#

（5）设置 vlan 1 端口 IP 地址，用于远程管理，此处省略了默认网关的配置。

switch# configure terminal
switch(config)#interface vlan 1
switch(config-if)#ip address 192.168.1.1 255.255.255.0
switch(config-if)#no shutdown
switch(config-if)#end
switch#write

## 六、实验测试和查看相关配置

（1）进入用户模式时的提示。

User Access Verification
Password:   #输入 console 端口的密码，同时没有字符显示
switch>

（2）进入特权模式时的提示。

switch>enable
Password:   #输入特权模式的密码，同时没有字符显示
switch #

（3）通过虚拟终端登录时的提示。

pc2#telnet 192.168.1.1
Trying 192.168.1.1 ... Open
User Access Verification
Password:   #输入虚拟终端的密码，同时没有字符显示
sw1>

（4）显示 console 端口状态信息。

Idle EXEC：空闲等待时间，如果超过空闲等待时间，会话会断开
sw1#show line console 0

| Tty Typ | Tx/Rx | A | Modem | Roty | AccO | AccI | Uses | Noise | Overruns | Int |

| * | 0 CTY | - | - | - | - | 0 | 1 | 0/0 | . |

Line 0, Location: "", Type: ""
Length: 24 lines, Width: 80 columns
Status: PSI Enabled, Ready, Active, Automore On
Capabilities: none
Modem state: Ready
Group codes:     0
Modem hardware state: noCTS noDSR    DTR RTS
Special Chars: Escape  Hold  Stop  Start  Disconnect  Activation
                ^^x    none   -     -      none
Timeouts:      Idle EXEC    Idle Session   Modem Answer  Session   Dispatch
               00:05:00     never                        00:10:00  not set
                            Idle Session Disconnect Warning
                            never
                            Login-sequence User Response
                            00:00:30
                            Autoselect Initial Wait
                            not set

Modem type is unknown    # Modem 的类型不知道
Session limit is not set    #会话限制没有设置
Time since activation: 00:03:19    #从登录到现在，已经过去的时间
Editing is enabled    #可以进行编辑
History is enabled, history size is 20    #可以记录历史命令，历史命令的条数为 20
DNS resolution in show commands is enabled    #可以进行 DNS 解析
Full user help is disabled
Allowed input transports are none
Allowed output transports are lat pad v120 lapb-ta telnet rlogin ssh
Preferred transport is lat
No output characters are padded
No special data dispatching characters

（5）显示虚拟终端状态信息。

sw1#sho line vty   0

| Tty Typ | Tx/Rx | A Modem | Roty AccO AccI | Uses | Noise | Overruns | Int |
| 130 VTY | - | - | - | 0 | 0 | 0/0 | - |

Line 130, Location: "", Type: ""
Length: 24 lines, Width: 80 columns
Baud rate (TX/RX) is 9600/9600
Status: No Exit Banner
Capabilities: none
Modem state: Idle
Group codes:     0
Special Chars: Escape  Hold  Stop  Start  Disconnect  Activation
                ^^x    none   -     -      none
Timeouts:      Idle EXEC    Idle Session   Modem Answer  Session   Dispatch
               00:05:00     never                        00:10:00  not set

```
                    Idle Session Disconnect Warning
                      never
                    Login-sequence User Response
                      00:00:30
                    Autoselect Initial Wait
                      not set
Modem type is unknown
Session limit is not set
Time since activation: never
Editing is enabled
History is enabled, history size is 20
DNS resolution in show commands is enabled
Full user help is disabled
Allowed input transports are lat pad v120 lapb-ta telnet rlogin ssh
Allowed output transports are lat pad v120 lapb-ta telnet rlogin ssh
Preferred transport is lat
No output characters are padded
No special data dispatching characters
```

## 七、注意问题

（1）在线路配置模式下，一定要配置命令 login。

（2）同时用 enable password xx 和 enable secret xx 命令设置密码的情况下，起作用的是 enable secret xx 命令设置的密码。

（3）通过 line vty 配置的虚拟终端登录，安全性很差，更多的时候是配置 ssh。

# 实验三 交换机的端口配置

理解交换机的三层与二层端口的区别，二层接口是工作在数据链路层的接口，无法设置 IP 地址，可以设置为 access 模式或者 trunk 模式；而三层端口可以设置 IP 地址，但是不能设置为 access 模式和 trunk 模式。三层交换机的端口可以在三层接口和二层接口之间切换，而纯二层接口或纯三层接口是不能在二层与三层之间切换的。

## 一、实验目的

熟悉交换机端口下的常用命令。

## 二、实验要内容

（1）对三层端口设置 IP 地址；
（2）设置端口速率；
（3）设置端口通信模式；
（4）查看端口状态。

## 三、实验需要掌握的命令

switch(config)#interface *interface-type slot/number*
！进入端口模式

switch(config-if)#switchport
！将交换机的端口从三层端口切换为二层端口（三层交换机的端口才有此命令）

switch(config-if)# no switchport
！将交换机的端口从二层端口切换为三层端口（三层交换机的端口才有此命令）

switch(config-if)#ip address *ip-address subnet-mask*
！设置端口 IP 地址（三层端口才可以设置 IP 地址）

switch(config-if)#no shutdown
！启用当前端口

switch(config-if)# shutdown
！关闭当前端口

switch(config-if)# speed *speed*
！设置端口的速率

switch(config-if)# duplex *duplex*
！设置端口的通信模式

switch(config)#interface range *interface-type slot/Start-number* – *end-number*
！同时对多个端口进行同样的配置（"–"的前后都有空格）

switch#write
！保存配置信息

switch #show interface FastEthernet *slot-number/number*
！显示指定端口信息
switch #show interface
！显示所有端口信息

### 四、实验拓扑图和网络文件

（1）网络拓扑图。

网络拓扑图如图 3-1 所示。

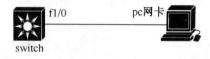

图 3-1　网络拓扑图

（2）网络文件。

网络文件 switch3.net 的内容如下：
```
# switch3.net 文件开始
[localhost]
    [[3660]]
    image = ..\c3660-is-mz.124-13.bin
    ram = 160
    [[ROUTER R1]]
    model = 3660
    slot1 = NM-16ESW      #加载模拟交换机模块
    f1/0 = NIO_gen_eth:\Device\NPF_{E2B31736-C716-4215-A9CB-ADA78C760F52}
# switch3.net 文件结束
```

### 五、实验配置步骤

（1）将端口 FastEthernet 1/0 切换为三层端口，接着设置端口 IP 地址、端口速率和端口通信模式，最后将端口启起。

```
Switch#configure terminal
Switch (config)#interface FastEthernet 1/0
Switch (config-if)#no switchport
Switch (config-if)#ip address    192.168.1.1 255.255.255.0
Switch (config-if)#speed 100
Switch (config-if)#duplex full
Switch (config-if)#no shutdown
Switch (config-if)#exit
```

（2）设置 4 个端口 FastEthernet 1/2 - 5，将四个端口切换为二层端口，接着设置端口速率和端口通信模式，然后启用 4 个端口，最后退出端口模式，进入特权模式，保存配置。

```
Switch (config)#interface range FastEthernet 1/2 - 5
Switch (config-if-range)#switchport
Switch (config-if-range)#speed 100
Switch (config-if-range)#duplex full
```

Switch (config-if-range)#no shutdown
Switch (config-if-range)#end
Switch #write

## 六、实验测试和查看相关配置

（1）在 PC 主机上 ping 192.168.1.1，看是否连通，结果如下：
C:\>ping 192.168.1.1

Pinging 192.168.1.1 with 32 bytes of data:

Reply from 192.168.1.1: bytes=32 time=38ms TTL=255
Reply from 192.168.1.1: bytes=32 time=43ms TTL=255
Reply from 192.168.1.1: bytes=32 time=54ms TTL=255
Reply from 192.168.1.1: bytes=32 time=7ms TTL=255

Ping statistics for 192.168.1.1:
　　Packets: Sent = 4, Received = 4, Lost = 0 (0% loss),
Approximate round trip times in milli-seconds:
Minimum = 7ms, Maximum = 54ms, Average = 35ms

（2）查看端口 interface FastEthernet 1/0 的状态信息。
查看端口状态信息时，主要关注下面的黑体字部分：
Switch #show interfaces FastEthernet 1/0
**FastEthernet1/0 is up, line protocol is up**　　#"FastEthernet1/0 is up"为端口已经启用；"line protocol is up"为端口的线路协议已经启用
　　Hardware is Fast Ethernet, address is cc00.0d94.f100 (bia cc00.0d94.f100)　　#端口的 MAC 地址
　　**Internet address is 192.168.7.159/24**　　#端口的 IP 地址
　　**MTU 1500 bytes, BW 100000 Kbit, DLY 100 usec**　　#MTU 为最大传输单元，BW 为端口带宽，DLY 为端口延时
　　reliability 255/255, txload 1/255, rxload 1/255
　　Encapsulation ARPA, loopback not set
　　Keepalive set (10 sec)
　　**Full-duplex, 100Mb/s**　　#传输模式为全双工
　　ARP type: ARPA, ARP Timeout 04:00:00
　　Last input 00:00:00, output never, output hang never
　　Last clearing of "show interface" counters never
　　Input queue: 0/75/0/0 (size/max/drops/flushes); Total output drops: 0
　　Queueing strategy: fifo
　　Output queue: 0/40 (size/max)
　　5 minute input rate 0 bits/sec, 0 packets/sec
　　5 minute output rate 0 bits/sec, 0 packets/sec
　　　　0 packets input, 0 bytes, 0 no buffer
　　　　Received 547 broadcasts, 0 runts, 0 giants, 0 throttles
　　　　0 input errors, 0 CRC, 0 frame, 0 overrun, 0 ignored
　　　　0 input packets with dribble condition detected
　　　　0 packets output, 0 bytes, 0 underruns
　　　　0 output errors, 0 collisions, 2 interface resets
　　　　0 babbles, 0 late collision, 0 deferred

            0 lost carrier, 0 no carrier
            0 output buffer failures, 0 output buffers swapped out
　　（3）查看端口 interface FastEthernet 1/2 的状态信息。
Switch #show interfaces FastEthernet 1/2
FastEthernet1/2 is up, line protocol is down
    Hardware is FastEthernet, address is cc00.0d94.f102 (bia cc00.0d94.f102)
    MTU 1500 bytes, BW 100000 Kbit, DLY 100 usec,
        reliability 255/255, txload 1/255, rxload 1/255
    Encapsulation ARPA, loopback not set
    Keepalive set (10 sec)
    Full-duplex, 100Mb/s
    ARP type: ARPA, ARP Timeout 04:00:00
    Last input never, output never, output hang never
    Last clearing of "show interface" counters never
    Input queue: 0/75/0/0 (size/max/drops/flushes); Total output drops: 0
    Queueing strategy: fifo
    Output queue: 0/40 (size/max)
    5 minute input rate 0 bits/sec, 0 packets/sec
    5 minute output rate 0 bits/sec, 0 packets/sec
        0 packets input, 0 bytes, 0 no buffer
        Received 0 broadcasts, 0 runts, 0 giants, 0 throttles
        0 input errors, 0 CRC, 0 frame, 0 overrun, 0 ignored
        0 input packets with dribble condition detected
        0 packets output, 0 bytes, 0 underruns
        0 output errors, 0 collisions, 2 interface resets
        0 babbles, 0 late collision, 0 deferred
        0 lost carrier, 0 no carrier
        0 output buffer failures, 0 output buffers swapped out

## 七、注意问题

（1）一对互相连接的端口要设置为速率相同，而且设置固定的速率，不要让端口自动匹配；

（2）一对互相连接的端口要设置为全双工或者半双工，而且不要让端口自动匹配；

（3）二层交换机的端口只能是二层端口，不能通过 no switchport 命令切换为三层端口。

# 实验四  VLAN 的配置

VLAN 的作用主要是通过软件在逻辑层面将一个物理设备划分为多个逻辑设备，或者将多个物理设备组合成一个逻辑设备；在不增加网络硬件设备的情况下，方便网络管理者管理网络和设置网络的安全。

## 一、实验目的

掌握 VLAN 的划分，将端口指定给一个 VLAN，设置端口模式及端口协议的封装。

## 二、实验内容

（1）掌握 VLAN 的划分；
（2）掌握设置端口的模式；
（3）掌握将端口指定给一个 VLAN；
（4）理解端口通道模式和中继模式的作用和区别。

## 三、实验原理

（1）VLAN 的概念。

IEEE 于 1999 年颁布了用以标准化 VLAN 实现方案的 802.1Q 协议标准草案。VLAN 技术的出现，使得管理员根据实际应用需求，把同一物理局域网内的不同用户逻辑地划分成不同的广播域，每一个 VLAN 都包含一组有着相同需求的计算机工作站，与物理上形成的 LAN 有着相同的属性。由于它是从逻辑上划分，而不是从物理上划分，所以同一个 VLAN 内的各个工作站没有限制在同一个物理范围中，即这些工作站可以在不同物理 LAN 网段。由 VLAN 的特点可知，一个 VLAN 内部的广播和单播流量都不会转发到其他 VLAN 中，从而有助于控制流量、减少设备投资、简化网络管理、提高网络的安全性。

交换技术的发展也加快了新的交换技术（VLAN）的应用速度。通过将企业网络划分为虚拟网络 VLAN 网段，可以强化网络管理和网络安全，控制不必要的数据广播。在共享网络中，一个物理的网段就是一个广播域。而在交换网络中，广播域可以是由一组任意选定的第二层网络地址（MAC 地址）组成的虚拟网段。这样，网络中工作组的划分可以突破共享网络中的地理位置限制，而完全根据管理功能来划分。这种基于工作流的分组模式，大大提高了网络规划和重组的管理功能。在同一个 VLAN 中的工作站，不论它们实际与哪个交换机连接，它们之间的通信就好像在独立的交换机上一样。同一个 VLAN 中的广播只有 VLAN 中的成员才能听到，而不会传输到其他的 VLAN 中去，这样可以很好地控制不必要的广播风暴产生。同时，若没有路由，不同 VLAN 之间不能相互通信，这样增加了企业网络中不同部门之间的安全性。网络管理员可以通过配置 VLAN 之间的路由来全面管理企业内部不同管理单元之间的信息互访。交换机是根据交换机的端口来划分 VLAN 的。所以，用户可以自由地在企业网络中移动办公，不论在何处接入交换网络，都可以与 VLAN 内其他用户自如通信。

VLAN 网络可以由混合的网络类型设备组成，比如：10M 以太网、100M 以太网、令牌网、

FDDI、CDDI 等，可以是工作站、服务器、集线器、网络上行主干等。

VLAN 除了能将网络划分为多个广播域，从而有效地控制广播风暴的发生，以及使网络的拓扑结构变得非常灵活的优点外，还可以用于控制网络中不同部门、不同站点之间的互相访问。VLAN 是为解决以太网的广播问题和安全性而提出的一种协议，它在以太网帧的基础上增加了 VLAN 头，用 VLAN ID 把用户划分为更小的工作组，限制不同工作组间的用户互访，每个工作组就是一个虚拟局域网。虚拟局域网的好处是可以限制广播范围，并能够形成虚拟工作组，动态管理网络。

（2）VLAN 的用途。

VLAN（Virtual Local Area Network，虚拟局域网）的用处非常多。通过认识 VLAN 的本质，将可以了解到其用处究竟在哪些地方。

第一，要知道 192.168.1.2/30 和 192.168.2.6/30 都属于不同的网段，都必须要通过路由器才能进行访问，凡是不同网段间要互相访问，都必须通过路由器。

第二，VLAN 的本质就是指一个网段，之所以叫做虚拟局域网，是因为它是在虚拟的路由器的接口下创建的网段。

下面给予说明。比如一个路由器只有一个用于终端连接的端口（当然这种情况基本不可能发生，只是为简化举例），这个端口被分配了 192.168.1.1/24 的地址。然而由于公司有两个部门，一个销售部，一个企划部，每个部门要求单独成为一个子网，有单独的服务器。那么当然可以划分为 192.168.1.0-127/25 和 192.168.1.128-255/25。但是路由器的物理端口只应该可以分配一个 IP 地址，那怎样来区分不同网段呢？这就可以在这个物理端口下，创建两个子接口，这样逻辑接口实现。

比如逻辑接口 F0/0.1 就分配 IP 地址 192.168.1.1/25，用于销售部，而 F0/0.2 就分配 IP 地址 192.168.1.129/25，用于企划部。这样就等于用一个物理端口确实现了两个逻辑接口的功能，这样就将原本只能划分一个网段的情形，扩展到了可以划分两个或者更多个网段的情形。这些网段因为是在逻辑接口下创建的，所以称之为虚拟局域网。这是在路由器的层次上阐述了 VLAN 的用处。

第三，将在交换机的层次上阐述 VLAN 的用处。

在现实中，由于很多原因必须划分出不同网段。比如，就简单地假设只有销售部和企划部两个网段。那么可以简单地将销售部全部接入一个交换机，然后接入路由器的一个端口，把企划部全部接入一个交换机，然后接入一个路由器端口。这种情况是 LAN。然而正如上面所说，如果路由器就是一个用于终端的接口，那么这两个交换机就必须接入同一个路由器的接口，这个时候，如果还想保持原来的网段划分，那么就必须使用路由器的子接口，创建 VLAN。

同样，比如两个交换机，如果想要每个交换机上的端口分别属于不同的网段，那么有几个网段，就提供几个路由器的接口，这时，虽然在路由器的物理接口上可以定义这个接口可以连接哪个网段，但是在交换机的层次上，它并不能区分哪个端口属于哪个网段，那么唯一实现能区分的方法就是划分 VLAN，使用 VLAN 就能区分出某个交换机端口的终端是属于哪个网段的。综上所述，当一个交换机上的所有端口中，有至少一个端口属于不同网段、路由器的一个物理端口要连接两个或者以上的网段时，就是 VLAN 发挥作用的时候，这就是 VLAN 的用处。

（3）VLAN 的优点。

1）广播风暴防范。

为限制网络上的广播，将网络划分为多个 VLAN，可减少参与广播风暴的设备数量。LAN

分段可以防止广播风暴波及整个网络。VLAN 可以提供建立防火墙的机制，防止交换网络的过量广播。使用 VLAN，可以将某个交换端口或用户赋予某一个特定的 VLAN 组，该 VLAN 组可以在一个交换网中或跨接多个交换机，在一个 VLAN 中的广播不会送到 VLAN 之外。同样，相邻的端口不会收到其他 VLAN 产生的广播。这样可以减少广播流量，释放带宽给用户应用，减少广播的产生。

2）安全。

增强局域网的安全性，含有敏感数据的用户组可与网络的其余部分隔离，从而降低泄露机密信息的可能性。不同 VLAN 内的报文在传输时是相互隔离的，即一个 VLAN 内的用户不能和其他 VLAN 内的用户直接通信，如果不同 VLAN 要进行通信，则需要通过路由器或三层交换机等三层设备。

3）降低成本。

由于成本高昂的网络升级需求减少，现有带宽和上行链路的利用率更高，因此可节约成本。

4）性能提高。

将第二层平面网络划分为多个逻辑工作组（广播域），可以减少网络上不必要的流量并提高性能。

5）提高 IT 员工效率。

VLAN 为网络管理带来了方便，因为有相似网络需求的用户将共享同一个 VLAN。

6）应用管理。

VLAN 将用户和网络设备聚合到一起，以支持商业需求或地域上的需求。通过职能划分，项目管理或特殊应用的处理都变得十分方便，例如可以轻松管理教师的电子教学开发平台。此外，也很容易确定升级网络服务的影响范围。

7）增加网络连接的灵活性。

借助 VLAN 技术，能将不同地点、不同网络、不同用户组合在一起，形成一个虚拟的网络环境，就像使用本地 LAN 一样方便、灵活、有效。VLAN 可以降低移动或变更工作站地理位置的管理费用，特别是一些业务情况有经常性变动的公司使用了 VLAN 后，这部分管理费用大大降低。

（4）组建 VLAN 的条件。

VLAN 是建立在物理网络基础上的一种逻辑子网,因此建立 VLAN 需要相应的支持 VLAN 技术的网络设备。当网络中的不同 VLAN 间进行相互通信时，需要路由的支持，这时就需要增加路由设备——要实现路由功能，既可采用路由器，也可采用三层交换机来完成。同时还严格限制了用户数量。

（5）VLAN 的划分。

1）根据端口来划分 VLAN。

许多 VLAN 厂商都利用交换机的端口来划分 VLAN 成员。被设定的端口都在同一个广播域中。例如，一个交换机的 1、2、3、4、5 端口被定义为虚拟网 AAA，同一交换机的 6、7、8 端口组成虚拟网 BBB。这样做允许各端口之间的通信，并允许共享型网络的升级。但是，这种划分模式将虚拟网限制在了一台交换机上。

第二代端口 VLAN 技术允许跨越多个交换机的多个不同端口划分 VLAN，不同交换机上的若干个端口可以组成同一个虚拟网。

以交换机端口来划分网络成员，其配置过程简单明了。因此，从目前来看，这种根据端口来划分 VLAN 的方式仍然是最常用的一种方式。

2）根据 MAC 地址划分 VLAN。

这种划分 VLAN 的方法是根据每个主机的 MAC 地址来划分，即对每个 MAC 地址的主机都配置它属于哪个组。这种划分 VLAN 的方法的最大优点就是当用户物理位置移动时，即从一个交换机换到其他的交换机时，VLAN 不用重新配置，所以，可以认为这种根据 MAC 地址划分的方法基于用户的 VLAN，这种方法的缺点是初始化时，所有的用户都必须进行配置，如果有几百个甚至上千个用户的话，配置是非常累的。而且这种划分方法也导致了交换机执行效率的降低，因为在每一个交换机的端口都可能存在很多个 VLAN 组的成员，这样就无法限制广播包了。另外，对于使用笔记本电脑的用户来说，其网卡可能经常更换，这样，VLAN 就必须不停地配置。

3）根据网络层划分 VLAN。

这种划分 VLAN 的方法是根据每个主机的网络层地址或协议类型（如果支持多协议）划分的，虽然这种划分方法是根据网络地址（比如 IP 地址），但它不是路由，与网络层的路由毫无关系。这种方法的优点是用户的物理位置改变了，不需要重新配置所属的 VLAN，而且可以根据协议类型来划分 VLAN，这对网络管理者来说很重要，而且，这种方法不需要附加的帧标签来识别 VLAN，这样可以减少网络的通信量。这种方法的缺点是效率低，因为检查每一个数据包的网络层地址是需要消耗处理时间的（相对于前面两种方法），一般的交换机芯片都可以自动检查网络上数据包的以太网帧头，但要让芯片能检查 IP 帧头则需要更高的技术，同时也更费时。当然，这与各个厂商的实现方法有关。

4）根据 IP 组播划分 VLAN。

IP 组播实际上也是一种 VLAN 的定义，即认为一个组播组就是一个 VLAN，这种划分的方法将 VLAN 扩大到了广域网，因此这种方法具有更大的灵活性，而且也很容易通过路由器进行扩展，当然这种方法不适合局域网，主要是效率不高。

5）根据规则划分 VLAN。

也称为根据策略划分 VLAN。这是最灵活的 VLAN 划分方法，具有自动配置的能力，能够把相关的用户连成一体，在逻辑划分上称为"关系网络"。网络管理员只需在网管软件中确定划分 VLAN 的规则（或属性），那么当一个站点加入网络中时，将会被"感知"，并被自动地包含进正确的 VLAN 中。同时，对站点的移动和改变也可自动识别和跟踪。采用这种方法，整个网络可以非常方便地通过路由器扩展网络规模。有的产品还支持一个端口上的主机分别属于不同的 VLAN，这在交换机与共享式 HUB 共存的环境中显得尤为重要。自动配置 VLAN 时，交换机中软件自动检查进入交换机端口的广播信息的 IP 源地址，然后软件自动将这个端口分配给一个由 IP 子网映射成的 VLAN。

6）根据用户划分 VLAN。

基于用户定义、非用户授权来划分 VLAN，是指为了适应特别的 VLAN 网络，根据具体的网络用户的特别要求来定义和设计 VLAN，而且可以让非 VLAN 群体用户访问 VLAN，但是需要提供用户密码，在得到 VLAN 管理的认证后才可以加入一个 VLAN。

以上划分 VLAN 的方式中，基于端口的 VLAN 端口方式建立在物理层上；MAC 方式建立在数据链路层上；网络层和 IP 广播方式建立在第三层上。

（6）VLAN 的标准。

对 VLAN 的标准，我们只是介绍两种比较通用的标准，当然也有一些公司具有自己的标准，比如 Cisco 公司的 ISL 标准，虽然不是一种大众化的标准，但是由于 Cisco Catalyst 交换机的大量使用，ISL 也成为一种不是标准的标准了。

1）802.10VLAN 标准。

在 1995 年，Cisco 公司提倡使用 IEEE 802.10 协议。在此之前，IEEE 802.10 曾经在全球范围内作为 VLAN 安全性的统一规范。Cisco 公司试图采用优化后的 802.10 帧格式在网络上传输 Frame Tagging 模式中所必须的 VLAN 标签。然而，大多数 802 委员会的成员都反对推广 802.10。因为，该协议是基于 Frame Tagging 方式的。

2）802.1Q。

在 1996 年 3 月，IEEE 802.1Internetworking 委员会结束了对 VLAN 初期标准的修订工作。新出台的标准进一步完善了 VLAN 的体系结构，统一了 Frame Tagging 方式中不同厂商的标签格式，并制定了 VLAN 标准在未来一段时间内的发展方向，形成的 802.1Q 的标准在业界获得了广泛的推广，并成为 VLAN 史上的一块里程碑。802.1Q 的出现打破了虚拟网依赖于单一厂商的僵局，从一个侧面推动了 VLAN 的迅速发展。另外，来自市场的压力使各大网络厂商立刻将新标准融合到他们各自的产品中。

3）Cisco ISL 标签。

ISL（Inter-Switch Link）是 Cisco 公司的专有封装方式，因此只能在 Cisco 的设备上支持。ISL 是一个在交换机之间、交换机与路由器之间及交换机与服务器之间传递多个 VLAN 信息及 VLAN 数据流的协议，通过在交换机直接的端口配置 ISL 封装，即可跨越交换机进行整个网络的 VLAN 分配和配置。

### 四、实验需要掌握的命令

switch#vlan database
！进入 VLAN 数据库设置
switch(vlan)#vlan *number*
！创建 VLAN *number*
switch(vlan)#no vlan *number*
！删除 VLAN *number*
switch(config)#interface FastEthernet *slot-number/number*
！进入指定端口
switch(config)# interface range FastEthernet *slot-number/number - number*
！进入指定的多个端口（端口号 number 和 "-" 之间有空格）
switch(config-if)#switchport
！将当前端口设置为二层端口（只在三层交换机才有效）
switch(config-if)#no switchport
！将当前端口设置为三层端口（只在三层交换机才有效）
switch(config-if)#switchport mode access
！将当前端口设为通道模式
switch(config-if)#switchport access vlan *number*
！将当前端口划分到一个 VLAN
switch(config-if)#switchport mode trunk

！将当前端口设为中继模式
switch(config-if)# switchport trunk encapsulation dot1q
！将当前中继模式端口的封装协议设为 dot1q
switch(config-if)# switchport trunk allowed vlan all
！将当前中继模式端口允许所用的 VLAN 通过
switch(config-if)# switchport trunk allowed vlan add *vlan_id*
！将当前中继模式端口添加 VLAN 号为 *vlan_id* 的 VLAN 通过
switch(config-if)# switchport trunk allowed vlan remove *vlan_id*
！将当前中继模式端口不允许 VLAN 号为 *vlan_id* 的 VLAN 通过
switch(config-if)# switchport trunk allowed vlan except *vlan_id*
！将当前中继模式端口除指定的 VLAN，其他的 VLAN 都可以通过
switch (config)#no ip routing
！关闭路由功能
switch(config-if)#switchport access vlan *number*
！将当前端口加入 vlan *number*

### 五、实验拓扑图和网络文件

（1）网络拓扑图。

网络拓扑图如图 4-1 所示。

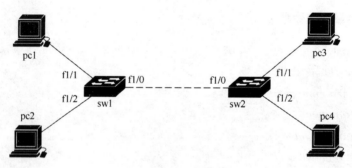

图 4-1　网络拓扑图

（2）网络文件。
port.net 文件内容
#port.net 文件开始
autostart = false
ghostios = true
sparsemem = true

[localhost]

　　[[3640]]
　　image = ..\C3640-IK.BIN
　　ram = 160

```
[[1720]]
image = ..\C1700-Y.BIN
ram = 16

[[router sw1]]
model = 3640
idlepc = 0x60577fb4          #降低模拟器对 CPU 的占用率
slot1 = NM-16ESW
f1/1 = pc1 f0
f1/2 = pc2 f0
f1/0 = sw2 f1/0

[[router sw2]]
model = 3640
idlepc = 0x60577fb4
slot1 = NM-16ESW
f1/1 = pc3 f0
f1/2 = pc4 f0

[[router pc1]]
model = 1720
idlepc = 0x8014e4ec

[[router pc2]]
model = 1720
idlepc = 0x8014e4ec

[[router pc3]]
model = 1720
idlepc = 0x8014e4ec

[[router pc4]]
model = 1720
idlepc = 0x8014e4ec
```
#port.net 文件结束

## 六、实验配置步骤

（1）sw1 的配置步骤。

```
switch>enable
switch#config terminal
switch(config)#hostname sw1
sw1(config)#no ip routing
sw1(config)#end
sw1#vlan database
sw1(vlan)#vlan 2
sw1(vlan)#vlan 3
sw1(vlan)#exit
```

```
sw1#config terminal
sw1(config)# interface FastEthernet 1/0
sw1(config-if)#switchport mode trunk
sw1(config-if)#duplex full
sw1(config-if)#speed 100
sw1(config-if)#switchport trunk encapsulation dot1q
sw1(config-if)#switchport trunk allowed vlan all
sw1(config-if)#no shutdown
sw1(config)# interface FastEthernet 1/1
sw1(config-if)#switchport mode access
sw1(config-if)#duplex full
sw1(config-if)#speed 100
sw1(config-if)#switchport access vlan 2
sw1(config-if)#no shutdown
sw1(config-if)# interface FastEthernet 1/2
sw1(config-if)#switchport mode access
sw1(config-if)#duplex full
sw1(config-if)#speed 100
sw1(config-if)#switchport access vlan 3
sw1(config-if)#no shutdown
sw1(config-if)#end
sw1#write
```

（2）sw2 配置步骤。

```
switch>enable
switch#config terminal
switch(config)#hostname sw2
sw2(config)#no ip routing
sw2(config)#end
sw2#vlan database
sw2(vlan)#vlan 2
sw2(vlan)#vlan 3
sw2(vlan)#exit
sw2#config terminal
sw2(config)# interface FastEthernet 1/0
sw2(config-if)#switchport mode trunk
sw2(config-if)#duplex full
sw2(config-if)#speed 100
sw2(config-if)#switchport trunk encapsulation dot1q
sw2(config-if)#switchport trunk allowed vlan all
sw2(config-if)#no shutdown
sw2(config)# interface FastEthernet 1/1
sw2(config-if)#switchport mode access
sw2(config-if)#duplex full
sw2(config-if)#speed 100
sw2(config-if)#switchport access vlan 2
sw2(config-if)#no shutdown
sw2(config-if)# interface FastEthernet 1/2
```

sw2(config-if)#switchport mode access
sw2(config-if)#duplex full
sw2(config-if)#speed 100
sw2(config-if)#switchport access vlan 3
sw2(config-if)#no shutdown
sw2(config-if)#end
sw2#write

(3) pc1 的配置步骤。
switch>enable
switch#config terminal
switch (config)#hostname pc1
pc1(config)#no ip routing
pc1(config)# interface FastEthernet 0
pc1(config-if)#ip address 192.168.1.3 255.255.255.0
pc1(config-if)#full-duplex
pc1(config-if)#no shutdown
pc1(config-if)#end
pc1#write

(4) pc2 的配置步骤。
switch>enable
switch#config terminal
switch (config)#hostname pc2
pc2(config)#no ip routing
pc2(config)# interface FastEthernet 0
pc2(config-if)#ip address 192.168.2.3 255.255.255.0
pc2(config-if)#full-duplex
pc2(config-if)#no shutdown
pc2(config-if)#end
pc2#write

(5) pc3 的配置步骤。
switch>enable
switch#config terminal
switch (config)#hostname pc3
pc3(config)#no ip routing
pc3(config)# interface FastEthernet 0
pc3(config-if)#ip address 192.168.1.4 255.255.255.0
pc3(config-if)#full-duplex
pc3(config-if)#no shutdown
pc3(config-if)#end
pc3#write

(6) pc4 的配置步骤。
switch>enable
switch#config terminal
switch (config)#hostname pc4
pc4(config)#no ip routing
pc4(config)# interface FastEthernet 0
pc4(config-if)#ip address 192.168.2.4 255.255.255.0

pc4(config-if)#full-duplex
pc4(config-if)#no shutdown
pc4(config-if)#end
pc4#write

### 七、实验测试和查看相关配置

（1）测试网络的连通性。

pc1 主机的地址是 192.168.1.3，pc2 主机的地址是 192.168.2.3，pc3 主机的地址是 192.168.1.4，pc4 主机的地址是 192.168.2.4，其中 pc1 和 pc3 属于 VLAN2，pc2 和 pc4 属于 VLAN3。在 pc1 主机上 ping 和主机在相同 VLAN 的主机是通的，ping 在不同 VLAN 的主机时不通。显示如下：

```
pc1#ping 192.168.1.4
Type escape sequence to abort.
Sending 5, 100-byte ICMP Echos to 192.168.1.4, timeout is 2 seconds:
!!!!!       #!表示 ping 通了对方
Success rate is 100 percent (5/5), round-trip min/avg/max = 28/57/104 ms


pc1#ping 192.168.2.4
Type escape sequence to abort.
Sending 5, 100-byte ICMP Echos to 192.168.2.4, timeout is 2 seconds:
.....       #.表示 ping 不通对方
Success rate is 0 percent (0/5)
```

（2）查看端口的 Trunk 状态信息。

Port：端口。

Mode：trunk 模式的状态，on 表示运行模式为 Trunk 模式。

Encapsulation：端口封装的协议。

Status：运行的状态，trunking 表示正在运行 Trunk 模式。

Native vlan：本地 VLAN。

sw1#show interfaces FastEthernet 1/0 trunk

| Port | Mode | Encapsulation | Status | Native vlan |
|---|---|---|---|---|
| Fa1/0 | on | 802.1q | trunking | 1 |

| Port | Vlans allowed on trunk | #允许那些 VLAN 信息通过此端口 |
|---|---|---|
| Fa1/0 | 1-1005 | |

| Port | Vlans allowed and active in management domain | #在管理域中允许和激活的 VLAN |
|---|---|---|
| Fa1/0 | 1-3 | |

Port    Vlans in spanning tree forwarding state and not pruned    #在生成树协议中处于转发状态的 VLAN，没有启用裁剪功能
Fa1/0    1-3

（3）查看交换端口的状态信息。

sw1#show interfaces FastEthernet 1/1 switchport
Name: Fa1/1     #端口名称

Switchport: Enabled     #二层的交换端口使能
Administrative Mode: static access   #管理员设置的模式为 Access 模式
Operational Mode: static access   #运行的模式为 Access 模式
Administrative Trunking Encapsulation: dot1q   #Trunk 模式封装的协议为 dot1q
Operational Trunking Encapsulation: native
Negotiation of Trunking: Disabled   #没有启用中继协商
Access Mode VLAN: 2 (VLAN0002)   #当前端口属于 VLAN2
Trunking Native Mode VLAN: 1 (default)   #中继模式的本地 VLAN 为 VLAN1
Trunking VLANs Enabled: ALL   #中继模式下默认允许所有 VLAN 信息通过
Trunking VLANs Active: 1   #激活的中继 VLAN 为 VLAN1
Priority for untagged frames: 0
Override vlan tag priority: FALSE
Voice VLAN: none
Appliance trust: none

（4）查看 VLAN 的状态信息。

sw1#show vlan-switch brief

| VLAN | Name | Status | Ports |
| --- | --- | --- | --- |
| 1 | default | active | Fa1/4, Fa1/5, Fa1/6, Fa1/7, Fa1/8, Fa1/3 |
|  |  |  | Fa1/9, Fa1/10, Fa1/11, Fa1/12, Fa1/13 |
|  |  |  | Fa1/14, Fa1/15 |
| 2 | VLAN0002 | active | Fa1/1 |
| 3 | VLAN0003 | active | Fa1/2 |
| 1002 | fddi-default | active |  |
| 1003 | token-ring-default | active |  |
| 1004 | fddinet-default | active |  |
| 1005 | trnet-default | active |  |

## 八、注意问题

（1）两个交换机 f1/0 连接的端口应该都配置相同的模式，本实验中应该都配置为 Trunk 模式，同时封装的协议也应该相同，最后还要让相应的 VLAN 通过此端口，否则网络会不通；

（2）对应的端口的速率和通信模式应该配置相同。

# 实验五　生成树协议的配置

在由交换机构成的交换网络中通常设计有冗余链路和设备。这种设计的目的是防止一个点的失败导致整个网络功能的丢失。虽然冗余设计可能消除单点失败问题，但也导致了交换回路的产生，它会带来如下问题：广播风暴、同一帧的多份拷贝、不稳定的 MAC 地址表。

因此，在交换网络中必须有一个机制来阻止回路，而生成树协议（Spanning Tree Protocol）的作用正在于此。

## 一、实验目的

掌握基于 VLAN 的生成树协议配置和应用。

## 二、实验内容

（1）熟悉生成树协议的原理和作用；
（2）掌握基于 VLAN 的生成树协议配置和应用。

## 三、实验原理

STP 协议是用来避免链路环路产生的广播风暴并提供链路冗余备份的协议。

STP 协议中定义了根桥（RootBridge）、根端口（RootPort）、指定端口（DesignatedPort）、路径开销（PathCost）等概念，目的就在于通过构造一棵自然树的方法达到裁剪冗余环路的目的，同时实现链路备份和路径最优化。用于构造这棵树的算法称为生成树算法（Spanning Tree Algorithm，SPA）。

要实现这些功能，网桥之间必须要进行一些信息的交流，这些信息交流单元就称为配置消息（Bridge Protocol Data Unit，BPDU）。STP BPDU 是一种二层报文，目的 MAC 是多播地址 01-80-C2-00-00-00，所有支持 STP 协议的网桥都会接收并处理收到的 BPDU 报文，该报文的数据区里携带了用于生成树计算的所有有用信息。

生成树协议的工作过程：

（1）选择根网桥：在全网中选择一个根网桥。

（2）比较网桥的 BID 值，值越小其优先级越高。ID 值是由两部分组成的：交换机的优先级和 MAC 地址，如果交换机的优先级相同则比较其 MAC 地址，地址值越小，其就被选举为根网桥。

（3）选择根端口：在每个非根交换机上选择根端口。

1）首先，比较根路径成本，根路径成本取决于链路的带宽，带宽越大，路径成本越低，则选该端口为根端口。

2）其次，如果根路径成本相同，则要比较所在对端交换机 BID 值，值越小，则其优先级越高。

3）最后，比较端口的 ID 值，该值分为两部分：端口优先级和端口编号，值小的被选为根端口。

（4）选择指定端口：在每条链路上选择一个指定端口，根网桥上所有端口都是指定端口。

1）首先，比较根路径成本。

2）其次，比较端口所在网桥的 ID 值。

3）最后，比较端口的 ID 值。

既不是根端口又不是指定端口的端口全部为阻塞状态。

STP 协议给透明网桥带来了新生。但是它还是有缺点的，STP 协议的缺陷主要表现在收敛速度上。

当拓扑发生变化，新的配置消息要经过一定的时延才能传播到整个网络，这个时延称为 Forward Delay，协议默认值是 15 秒。在所有网桥收到这个变化的消息之前，若旧拓扑结构中处于转发的端口还没有发现自己应该在新的拓扑中停止转发，则可能存在临时环路。为了解决临时环路的问题，生成树使用了一种定时器策略，即在端口从阻塞状态到转发状态的中间加上一个只学习 MAC 地址但不参与转发的中间状态，两次状态切换的时间长度都是 Forward Delay，这样就可以保证在拓扑变化的时候不会产生临时环路。但是，这个看似良好的解决方案实际上带来的却是至少两倍 Forward Delay 的收敛时间。

### 四、实验需要掌握的命令

switch(config)#spanning-tree vlan *vlan-number*
！启用 stp 生成树（基于 VLAN）

switch(config)#spanning-tree vlan *vlan-number* root primary
！指定根交换机（基于 VLAN）

switch(config)#spanning-tree vlan *vlan-number* root secondary
！指定备用根交换机（基于 VLAN）

switch(config)#spanning-tree vlan *vlan-number* priority *number*
！指定交换机优先级（基于 VLAN），*number* 是交换机优先级，其值越小，则优先级越高，在 0～65535 之间

switch(config)#no spanning-tree vlan *vlan-number* priority
！将交换机的优先级恢复默认值（基于 VLAN）

switch(config-if)#spanning-tree vlan *vlan-number* cost *number*
！指定端口成本（起用 Trunk 的端口模式下），*number* 是端口的费用，其值在 1～65535 之间

switch(config-if)#spanning-tree vlan *vlan-number* port-prioty *number*
！指定交换机端口优先级（基于 VLAN），*number* 是端口优先级，其值越小，则优先级越高，其值的大小必须是 8 的整数倍，在 0～255 之间

switch(config-if)#spanning-tree portfast
！配置速端口（连接终端设备的端口状态），如 PC 机

switch(config)#spanning-tree uplinkfast [max-update-rate *number* ]
！配置上行速端口，*number* 为每秒更新包的最大值，其值为 0～65535

switch(config)#spanning-tree vlan *vlan-number* hello-time *time*
！配置交换机 hello 时间（基于 VLAN），*time* 的值在 1～10 秒

switch(config)#spanning-tree vlan *vlan-number* forward-time *time*

！修改转发延迟计时器（基于 VLAN），*time* 的值在 4～30 秒
switch(config)#spanning-tree vlan *vlan-number* max-age *time*
！修改最大老化时间（基于 VLAN），*time* 的值在 6～40 秒
switch#show spanning-tree brief
！显示生成树配置信息
switch#show spanning-tree vlan brief
！显示详细 VLAN 生成树配置信息
switch#show spanning-tree interface *type interface-number* brief
！显示详细 VLAN 生成树端口配置信息

### 五、实验拓扑图和网络文件

（1）网络拓扑图。

网络拓扑图如图 5-1 所示。

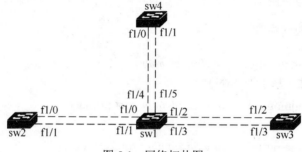

图 5-1　网络拓扑图

（2）网络文件。

```
autostart = false
ghostios = true
sparsemem = true

[localhost]

    [[3745]]
    image = ..\c3745-adventerprisek9-mz.124-16.bin
    ram = 160

    [[ROUTER sw1]]
    model = 3745
    slot1 = NM-16ESW
    f1/0 = sw2 f1/0
    f1/1 = sw2 f1/1
    f1/2 = sw3 f1/2
    f1/3 = sw3 f1/3

    [[ROUTER sw2]]
    model = 3745
```

slot1 = NM-16ESW

[[ROUTER sw3]]
model = 3745
slot1 = NM-16ESW

[[ROUTER sw4]]
model = 3745
slot1 = NM-16ESW
f1/0 = sw1 f1/4
f1/1 = sw1 f1/5

### 六、实验配置步骤

（1）交换机 sw1 的配置。

switch>enable
switch #config terminal
switch(config)# hostname sw1
sw1(config)#exit
sw1#vlan database
sw1(vlan)#vlan 2
sw1(vlan)#vlan 3
sw1(vlan)#exit
sw1#config terminal
sw1(config)# spanning-tree vlan 2
sw1(config)# spanning-tree vlan 3
sw1(config)# spanning-tree vlan 2 root primary
sw1(config)# spanning-tree vlan 3 root secondary
sw1(config)# interface range f1/0 - 5
sw1(config-if -range)# switchport mode trunk
sw1(config-if -range)# switchport trunk encapsulation dot1q
sw1(config-if -range)# switchport trunk allowed vlan all
sw1(config-if -range)#speed 100
sw1(config-if -range)#duplex full
sw1(config-if -range)#no shutdown
sw1(config-if -range)#exit
sw1(config)# interface range f1/1 , f1/3 , f1/5
sw1(config-if -range)# spanning-tree vlan 3 cost 16
sw1(config-if -range)#end
sw1#write

（2）交换机 sw2 的配置。

sw2>enable
sw2#vlan database
sw2(vlan)#vlan 2
sw2(vlan)#vlan 3
sw2(vlan)#exit
sw2#config terminal
sw2(config)# spanning-tree vlan 2

```
sw2(config)# spanning-tree vlan 3
sw2(config)# interface range f1/0 - 1
sw2(config-if-range)# switchport mode trunk
sw2(config-if-range)# switchport trunk encapsulation dot1q
sw2(config-if-range)# switchport trunk allowed vlan all
sw2(config-if-range)#speed 100
sw2(config-if-range)#duplex full
sw2(config-if-range)#no shutdown
sw2(config-if-range)#exit
sw2(config-if)# interface f1/1
sw2(config-if)# spanning-tree vlan 3 cost 16
sw2(config-if)#end
sw2#write
```

（3）交换机 sw3 的配置。

```
sw3>enable
sw3#vlan database
sw3(vlan)#vlan 2
sw3(vlan)#vlan 3
sw3(vlan)#exit
sw3#config terminal
sw3(config)# spanning-tree vlan 2
sw3(config)# spanning-tree vlan 3
sw3(config)# interface range f1/2 - 3
sw3(config-if-range)# switchport mode trunk
sw3(config-if-range)# switchport trunk encapsulation dot1q
sw3(config-if-range)# switchport trunk allowed vlan all
sw3(config-if-range)#speed 100
sw3(config-if-range)#duplex full
sw3(config-if-range)#no shutdown
sw3(config-if-range)#exit
sw3(config)# interface f1/3
sw3(config-if)# spanning-tree vlan 3 cost 16
sw3(config-if)#end
sw3#write
```

（4）交换机 sw4 的配置。

```
sw4>enable
sw4#vlan database
sw4(vlan)#vlan 2
sw4(vlan)#vlan 3
sw4(vlan)#exit
sw4#config terminal
sw4(config)# spanning-tree vlan 2
sw4(config)# spanning-tree vlan 3
sw4(config)# spanning-tree vlan 3 root primary
sw4(config)# spanning-tree vlan 2 root secondary
sw4(config)# interface range f1/0 - 1
sw4(config-if-range)# switchport mode trunk
```

sw4(config-if -range)# switchport trunk encapsulation dot1q
sw4(config-if -range)# switchport trunk allowed vlan all
sw4(config-if -range)#speed 100
sw4(config-if -range)#duplex full
sw4(config-if -range)#no shutdown
sw4(config-if -range)#exit
sw4(config-if)# interface f1/1
sw4(config-if)# spanning-tree vlan 3 cost 16
sw4(config-if)#end
sw4#write

## 七、实验测试和查看相关配置

（1）显示交换机 sw1 的生成树协议的配置信息。

Port ID：表示端口 ID 值。

Prio：端口优先级。

Cost：端口到根桥的路径成本值。

Sts：端口状态，FWD 为转发状态，BLK 为阻塞状态。

Bridge ID：如果当前网桥不是根桥，则为与此端口连接的网桥的 ID 值；否则为根网桥的 ID 值。

sw1#show spanning-tree brief

VLAN1
  Spanning tree enabled protocol ieee
  Root ID    Priority    32768       #根桥的优先级
             Address     c400.11a8.0000    #根桥的 MAC 地址
             This bridge is the root    #表示当前网桥就是根桥
             Hello Time   2 sec   Max Age 20 sec   Forward Delay 15 sec

  Bridge ID  Priority    32768       #当前网桥的优先级
             Address     c400.11a8.0000    #当前根桥的 MAC 地址
             Hello Time   2 sec   Max Age 20 sec   Forward Delay 15 sec
             Aging Time 300

| Interface Name | Port ID | Prio Cost | Sts Cost | Designated Bridge ID | Port ID |
|---|---|---|---|---|---|
| FastEthernet1/0 | 128.41 | 128 | 19 FWD | 0 32768 c400.11a8.0000 | 128.41 |
| FastEthernet1/1 | 128.42 | 128 | 19 FWD | 0 32768 c400.11a8.0000 | 128.42 |
| FastEthernet1/2 | 128.43 | 128 | 19 FWD | 0 32768 c400.11a8.0000 | 128.43 |
| FastEthernet1/3 | 128.44 | 128 | 19 FWD | 0 32768 c400.11a8.0000 | 128.44 |
| FastEthernet1/4 | 128.45 | 128 | 19 FWD | 0 32768 c400.11a8.0000 | 128.45 |
| FastEthernet1/5 | 128.46 | 128 | 19 FWD | 0 32768 c400.11a8.0000 | 128.46 |

VLAN2
  Spanning tree enabled protocol ieee

```
       Root ID    Priority     8192
                  Address      c400.11a8.0001
                  This bridge is the root
                  Hello Time   2 sec   Max Age 20 sec   Forward Delay 15 sec

       Bridge ID  Priority     8192
                  Address      c400.11a8.0001
                  Hello Time   2 sec   Max Age 20 sec   Forward Delay 15 sec
                  Aging Time 300

    Interface                                    Designated
       Name       Port ID  Prio Cost  Sts Cost         Bridge ID           Port ID
    -------------------- ------- ---- ----- --- ----- -------------------- -------
    FastEthernet1/0      128.41    128    19 FWD    0    8192 c400.11a8.0001   128.41
    FastEthernet1/1      128.42    128    19 FWD    0    8192 c400.11a8.0001   128.42
    FastEthernet1/2      128.43    128    19 FWD    0    8192 c400.11a8.0001   128.43
    FastEthernet1/3      128.44    128    19 FWD    0    8192 c400.11a8.0001   128.44
    FastEthernet1/4      128.45    128    19 FWD    0    8192 c400.11a8.0001   128.45
    FastEthernet1/5      128.46    128    19 FWD    0    8192 c400.11a8.0001   128.46

    VLAN3
       Spanning tree enabled protocol ieee
       Root ID    Priority     8192
                  Address      c403.11a8.0002
                  Cost         16
                  Port         46 (FastEthernet1/5)
                  Hello Time   2 sec   Max Age 20 sec   Forward Delay 15 sec

       Bridge ID  Priority     16384
                  Address      c400.11a8.0002
                  Hello Time   2 sec   Max Age 20 sec   Forward Delay 15 sec
                  Aging Time 300

    Interface                                    Designated
       Name       Port ID  Prio Cost  Sts Cost         Bridge ID           Port ID
    -------------------- ------- ---- ----- --- ----- -------------------- -------
    FastEthernet1/0      128.41    128    19 FWD   16   16384 c400.11a8.0002   128.41
    FastEthernet1/1      128.42    128    16 FWD   16   16384 c400.11a8.0002   128.42
    FastEthernet1/2      128.43    128    19 FWD   16   16384 c400.11a8.0002   128.43
    FastEthernet1/3      128.44    128    16 FWD   16   16384 c400.11a8.0002   128.44
    FastEthernet1/4      128.45    128    19 BLK    0    8192 c403.11a8.0002   128.41
    FastEthernet1/5      128.46    128    16 FWD    0    8192 c403.11a8.0002   128.42
```

（2）显示交换机 sw2 的生成树协议的配置信息。

```
sw2#show spanning-tree brief

VLAN1
    Spanning tree enabled protocol ieee
```

```
Root ID      Priority     32768      #根桥的优先级
             Address      c400.11a8.0000    #根桥的 MAC 地址
             Cost         19    #到根桥的路径成本值
             Port         41 (FastEthernet1/0)    #根端口编号
             Hello Time   2 sec    Max Age 20 sec    Forward Delay 15 sec

Bridge ID    Priority     32768    #当前网桥的优先级
             Address      c401.11a8.0000    #当前根桥的 MAC 地址
             Hello Time   2 sec    Max Age 20 sec    Forward Delay 15 sec
             Aging Time 300
```

| Interface Name | Port ID | Prio | Cost | Sts Cost | Designated Bridge ID | Port ID |
|---|---|---|---|---|---|---|
| FastEthernet1/0 | 128.41 | 128 | | 19 FWD | 0 32768 c400.11a8.0000 | 128.41 |
| FastEthernet1/1 | 128.42 | 128 | | 19 BLK | 0 32768 c400.11a8.0000 | 128.42 |

VLAN2
```
Spanning tree enabled protocol ieee
Root ID      Priority     8192
             Address      c400.11a8.0001
             Cost         19
             Port         41 (FastEthernet1/0)
             Hello Time   2 sec    Max Age 20 sec    Forward Delay 15 sec

Bridge ID    Priority     32768
             Address      c401.11a8.0001
             Hello Time   2 sec    Max Age 20 sec    Forward Delay 15 sec
             Aging Time 300
```

| Interface Name | Port ID | Prio | Cost | Sts Cost | Designated Bridge ID | Port ID |
|---|---|---|---|---|---|---|
| FastEthernet1/0 | 128.41 | 128 | | 19 FWD | 0  8192 c400.11a8.0001 | 128.41 |
| FastEthernet1/1 | 64.42  | 64  | | 19 BLK | 0  8192 c400.11a8.0001 | 128.42 |

VLAN3
```
Spanning tree enabled protocol ieee
Root ID      Priority     8192
             Address      c403.11a8.0002
             Cost         32
             Port         42 (FastEthernet1/1)
             Hello Time   2 sec    Max Age 20 sec    Forward Delay 15 sec

Bridge ID    Priority     32768
             Address      c401.11a8.0002
             Hello Time   2 sec    Max Age 20 sec    Forward Delay 15 sec
             Aging Time 300
```

| Interface Name | Port ID | Prio Cost | Sts Cost | Designated Bridge ID | Port ID |
|---|---|---|---|---|---|
| FastEthernet1/0 | 128.41 | 128 | 19 BLK | 16 16384 c400.11a8.0002 | 128.41 |
| FastEthernet1/1 | 128.42 | 128 | 16 FWD | 16 16384 c400.11a8.0002 | 128.42 |

（3）显示交换机 sw3 的生成树协议的配置信息。

sw3#show spanning-tree brief

VLAN1
  Spanning tree enabled protocol ieee
  Root ID    Priority    32768
              Address     c400.11a8.0000
              Cost        19
              Port        43 (FastEthernet1/2)
              Hello Time   2 sec   Max Age 20 sec   Forward Delay 15 sec

  Bridge ID   Priority    32768
              Address     c402.11a8.0000
              Hello Time   2 sec   Max Age 20 sec   Forward Delay 15 sec
              Aging Time 300

| Interface Name | Port ID | Prio Cost | Sts Cost | Designated Bridge ID | Port ID |
|---|---|---|---|---|---|
| FastEthernet1/2 | 128.43 | 128 | 19 FWD | 0 32768 c400.11a8.0000 | 128.43 |
| FastEthernet1/3 | 128.44 | 128 | 19 BLK | 0 32768 c400.11a8.0000 | 128.44 |

VLAN2
  Spanning tree enabled protocol ieee
  Root ID    Priority    8192
              Address     c400.11a8.0001
              Cost        19
              Port        43 (FastEthernet1/2)
              Hello Time   2 sec   Max Age 20 sec   Forward Delay 15 sec

  Bridge ID   Priority    32768
              Address     c402.11a8.0001
              Hello Time   2 sec   Max Age 20 sec   Forward Delay 15 sec
              Aging Time 300

| Interface Name | Port ID | Prio Cost | Sts Cost | Designated Bridge ID | Port ID |
|---|---|---|---|---|---|
| FastEthernet1/2 | 128.43 | 128 | 19 FWD | 0 8192 c400.11a8.0001 | 128.43 |
| FastEthernet1/3 | 128.44 | 128 | 19 BLK | 0 8192 c400.11a8.0001 | 128.44 |

VLAN3
   Spanning tree enabled protocol ieee
   Root ID    Priority    8192
               Address     c403.11a8.0002
               Cost        32
               Port        44 (FastEthernet1/3)
               Hello Time  2 sec  Max Age 20 sec  Forward Delay 15 sec

   Bridge ID  Priority    32768
               Address     c402.11a8.0002
               Hello Time  2 sec  Max Age 20 sec  Forward Delay 15 sec
               Aging Time 300

```
Interface                                   Designated
  Name          Port ID  Prio Cost  Sts Cost        Bridge ID         Port ID
-------------------- ------- ---- ----- --- ----- --------------------- -------
FastEthernet1/2    128.43    128     19 BLK    16 16384 c400.11a8.0002  128.43
FastEthernet1/3    128.44    128     16 FWD    16 16384 c400.11a8.0002  128.44
```

（4）显示交换机 sw4 的生成树协议的配置信息。

sw4#show spanning-tree brief
VLAN1
   Spanning tree enabled protocol ieee
   Root ID    Priority    32768
               Address     c400.11a8.0000
               Cost        19
               Port        41 (FastEthernet1/0)
               Hello Time  2 sec  Max Age 20 sec  Forward Delay 15 sec

   Bridge ID  Priority    32768
               Address     c403.11a8.0000
               Hello Time  2 sec  Max Age 20 sec  Forward Delay 15 sec
               Aging Time 300

```
Interface                                   Designated
  Name          Port ID  Prio Cost  Sts Cost        Bridge ID         Port ID
-------------------- ------- ---- ----- --- ----- --------------------- -------
FastEthernet1/0    128.41    128     19 FWD    0 32768 c400.11a8.0000   128.45
FastEthernet1/1    128.42    128     19 BLK    0 32768 c400.11a8.0000   128.46
```

VLAN2
   Spanning tree enabled protocol ieee
   Root ID    Priority    8192
               Address     c400.11a8.0001
               Cost        19
               Port        41 (FastEthernet1/0)
               Hello Time  2 sec  Max Age 20 sec  Forward Delay 15 sec

```
        Bridge ID   Priority    16384
                    Address     c403.11a8.0001
                    Hello Time   2 sec   Max Age 20 sec   Forward Delay 15 sec
                    Aging Time 300

    Interface                                Designated
       Name          Port ID  Prio Cost  Sts Cost        Bridge ID         Port ID
    -------------------- -------- ---- ----- --- ----- -------------------- -------
    FastEthernet1/0     128.41    128    19 FWD    0   8192 c400.11a8.0001   128.45
    FastEthernet1/1     128.42    128    19 BLK    0   8192 c400.11a8.0001   128.46

VLAN3
    Spanning tree enabled protocol ieee
    Root ID     Priority    8192
                Address     c403.11a8.0002
                This bridge is the root
                Hello Time   2 sec   Max Age 20 sec   Forward Delay 15 sec

    Bridge ID   Priority    8192
                Address     c403.11a8.0002
                Hello Time   2 sec   Max Age 20 sec   Forward Delay 15 sec
                Aging Time 300

    Interface                                Designated
       Name          Port ID  Prio Cost  Sts Cost        Bridge ID         Port ID
    -------------------- -------- ---- ----- --- ----- -------------------- -------
    FastEthernet1/0     128.41    128    19 FWD    0   8192 c403.11a8.0002   128.41
    FastEthernet1/1     128.42    128    16 FWD    0   8192 c403.11a8.0002   128.42
```

从上面的显示信息可以看出,交换机对于每一个 VLAN 都是一个树型结构,没有环路;同时,不同的 VLAN 经过的路径不同,从而起到了负载均衡的作用。

## 八、注意问题

(1)在设置端口优先级时要注意,其数值应该是 8 的整数倍,否则无法设置,同时交换机会报错。

(2)如果确定某个端口连接的是终端设备,可以用 switch(config-if)#spanning-tree portfast 命令将此端口设置为 fast 端口,则生成树协议的收敛速度会加快。

# 实验六　VTP 的配置

VLAN 中继协议利用第二层中继帧，在一组交换机之间进行 VLAN 信息通信。VTP 从一个中心控制点开始，维护整个企业网上 VLAN 的添加和重命名工作，确保配置的一致性。

## 一、实验目的

掌握 VTP 协议的配置与调试。

## 二、实验内容

（1）熟练理解相关命令的作用；
（2）掌握 VTP 协议的配置与调试；
（3）查看 VTP 协议的相关信息。

## 三、实验原理

（1）VTP 概述和工作原理。

VTP 是一种消息协议，使用第 2 层帧，在全网的基础上管理 VLAN 的添加、删除和重命名，以实现 VLAN 配置的一致性。可以用 VTP 管理网络中 VLAN1 到 1005。

有了 VTP，就可以在 VTP 域的 Server 模式的交换机上集中管理 VLAN 的相关配置，所做的变更会被自动传播到网络中所有其他的交换机上（前提是在同一个 VTP 域）。为了实现此功能，必须先建立一个 VTP 管理域，以使它能管理网络上当前的 VLAN。在同一管理域中的交换机共享其 VLAN 信息，并且，一个交换机只能参加到一个 VTP 管理域，不同域中的交换机不能共享 VTP 信息。

交换机间交换下列信息：
1）管理域域名。
2）配置的修订号。
3）已知虚拟局域网的配置信息。

交换机使用配置修订号来决定当前交换机的内部数据是否应该接受从其他交换机发来的 VTP 更新信息。如果接收到的 VTP 更新配置修订号与内部数据库的修订号相同域者比它小，交换机忽略更新。否则，就更新内部数据库，接受更新信息。

VTP 管理域在安全模式下，必须配置一个在 VTP 域中所有交换机唯一的口令。

VTP 的运行有如下特点：
1）VTP 通过发送到特定 MAC 地址 01-00-0C-CC-CC-CC 的组播 VTP 消息进行工作。
2）VTP 通告只通过中继端口传递。
3）VTP 消息通过 VLAN1 传送（这就是不能将 VLAN1 从中继链路中去除的原因）。
4）在经过了 DTP（Dynamic Trunk Protocol，动态中继协议）自动协商，启动了中继之后，VTP 信息就可以沿着中继链路传送。
5）VTP 域内的每台交换机都定期在每个中继端口上发送通告到保留的 VTP 组播地址，VTP 通告可以封装在 ISL 或者 IEEE 802.1Q 帧内。

（2）VTP 域。

VTP 域也称为 VLAN 管理域，由一个以上共享 VTP 域名的相互接连的交换机组成。

要使用 VTP，就必须为每台交换机指定 VTP 域名。VTP 信息只能在 VTP 域内保持。一台交换机可属于并且只属于一个 VTP 域。

默认情况下，CATALYST 交换机处于 VTP 服务器模式，并且不属于任何管理域，直到交换机通过中继链路接收了关于一个域的通告，或者在交换机上配置了一个 VLAN 管理域，交换机才能在 VTP 服务器上，把创建或者更改 VLAN 的消息通告给本管理域内的其他交换机。如果在 VTP 服务器上进行了 VLAN 配置变更，所做的修改会传播到 VTP 域内的所有交换机上。如果交换机配置为"透明"模式，可以创建或者修改 VLAN，但所做的修改只影响单个的交换机。控制 VTP 功能的一项关键参数是 VTP 配置修改编号，这个 32 位的数字表明了 VTP 配置的特定修改版本。配置修改编号的取值从 0 开始，每修改一次，就增加 1，直到达到 4294967295，然后循环归 0，并重新开始增加。每个 VTP 设备会记录自己的 VTP 配置修改编号；VTP 数据包会包含发送者的 VTP 配置修改编号，这一信息用于确定接收到的信息是否比当前的信息更新。要将交换机的配置修改号置为 0，只需要禁用中继，改变 VTP 的名称，然后再次启用中继。

VTP 域的要求如下：

1）域内的每台交换机必须使用相同的 VTP 域名，不论是通过配置实现，还是由交换机自动实现。

2）CATALYST 交换机必须是相邻的，这意味着 VTP 域内的所有交换机形成了一颗相互连接的树。每台交换机都通过这棵树与其他交换机交换信息。

3）在所有的交换机之间，必须启用中继。

（3）VTP 的运行模式。

VTP 模式有 3 种：

1）服务器模式（SERVER，默认）。

VTP 服务器控制着它们所在域中 VLAN 的生成和修改。所有的 VTP 信息都被通告在本域中的其他交换机上，而且，所有这些 VTP 信息都是被其他交换机同步接收的。

2）客户机模式（CLIENT）。

VTP 客户机不允许管理员创建、修改或删除 VLAN。它们监听本域中其他交换机的 VTP 通告，并相应地修改它们的 VTP 配置情况。

3）透明模式（TRANSPARENT）。

VTP 透明模式中的交换机不参与 VTP。当交换机处于透明模式时，它不通告其 VLAN 配置信息。而且，它的 VLAN 数据库更新与收到的通告也不保持同步。但它可以创建和删除本地的 VLAN。不过，这些 VLAN 的变更不会传播到其他任何交换机上。

四、实验需要掌握的命令

switch(vlan)#vtp domain
#设置 VTP 域名

switch(vlan)#vtp password
#设置 VTP 密码

switch(vlan)#vtp server
#设置 VTP 模式为服务器模式

switch(vlan)#vtp client
#设置 VTP 模式为客户端模式

switch(vlan)#vtp transparent
#设置 VTP 模式为透明模式

switch(vlan)#vtp pruning
#开启 VTP 的裁剪功能

switch(config-if)#switchport trunk allowed vlan all
#允许所有的 VLAN 信息通过此 Trunk 端口

switch(config-if)#switchport trunk allowed vlan {add|remove|except *vlan_ID*}
#对通过此 Trunk 端口的 VLAN 信息进行裁剪，add 是增加 VLAN 号为 *vlan_ID* 通过此 Trunk 端口；remove 是删除 VLAN 号为 *vlan_ID* 通过此 Trunk 端口；except 是除 VLAN 号为 *vlan_ID* 通过此 Trunk 端口，其他的端口都删除（不包含默认 VLAN）。*vlan_ID* 可以是多个 VLAN 号列表，如 2,3,4 与 2-4 的作用相同，这两种用法可以组合应用，如 2,4-6。

switch #show vtp status
#查看 VTP 配置信息

switch #show vlan-switch [brief|ID *vlan_ID*|name *vlan_name*]
#显示 VLAN 的配置信息，*vlan_ID* 是 VLAN 号，*vlan_name* 是 VLAN 的名称

switch #show interface FastEthernet *slot-number/number - number*
#查看指定端口信息

switch(config)#ip default-gateway *ip-address*
#设置默认网关（此命令在二层交换机下才有效，如果是在三层交换机使用此命令，必须使用 no ip routing 命令关闭路由功能，否则无效）

### 五、实验拓扑图和网络文件

（1）网络拓扑图。

网络拓扑图如图 6-1 所示。

（2）网络文件。

```
#vtp.net START
autostart = false
ghostios = true
sparsemem = true


[localhost]
    [[3660]]
        image = ..\C3660-IS.BIN
        ram = 160

    [[3640]]
        image = ..\C3640-IK.BIN
        ram = 160

    [[1720]]
```

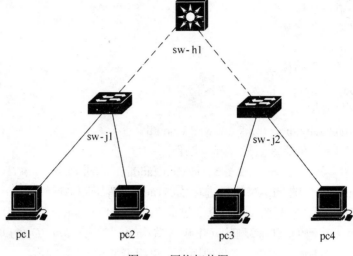

图 6-1 网络拓扑图

image = ..\C1700-Y.BIN
ram = 16

[[router sw-h1]]
 model = 3660
 idlepc = 0x6064b000
 slot1 = NM-16ESW
 f1/1 = sw-j1 f1/0
 f1/2 = sw-j2 f1/0

[[router sw-j1]]
 model = 3640
 idlepc = 0x6050c860
 slot1 = NM-16ESW
 f1/3 = pc2 f0
 f1/6 = pc2 f0

[[router sw-j2]]
 model = 3640
 idlepc = 0x6050c860
 slot1 = NM-16ESW
 f1/3 = pc3 f0
 f1/6 = pc4 f0

[[router pc1]]
 model = 1720

[[router pc2]]
 model = 1720

[[router pc3]]

```
            model = 1720

        [[router pc4]]
            model = 1720
#vtp.net END
```

## 六、实验配置步骤

（1）交换机 sw-h1 的配置。

```
switch>enable
switch#config terminal
switch(config)# hostname sw-h1
sw-h1(config)#exit
sw-h1#vlan database
sw-h1(vlan)#vlan 2
sw-h1(vlan)#vlan 3
sw-h1(vlan)#vlan 4
sw-h1(vlan)#vtp domain test
sw-h1(vlan)#vtp server
sw-h1(vlan)#vtp password abc
sw-h1(vlan)#vtp pruning
sw-h1(vlan)#vtp v2-mode
sw-h1(vlan)#exit
sw-h1#config terminal
sw-h1(config)# interface range f1/1 - 2
sw-h1(config-if-range)#switchport mode trunk
sw-h1(config-if-range)#switchport trunk encapsulation dot1q
sw-h1(config-if-range)#switchport trunk allowed vlan all
sw-h1(config-if-range)#speed 100
sw-h1(config-if-range)#duplex full
sw-h1(config-if-range)#no shutdown
sw-h1(config-if-range)#end
sw-h1#write
```

（2）交换机 sw-j1 的配置。

```
switch>enable
switch#config terminal
switch(config)# hostname sw-j1
sw-j1(config)#exit
sw-j1#vlan database
sw-j1(vlan)#vtp domain test
sw-j1(vlan)#vtp client
sw-j1(vlan)#vtp password abc
sw-j1(vlan)#exit
sw-j1#config terminal
sw-j1(config)# interface f1/0
sw-j1(config-if)#switchport mode trunk
sw-j1(config-if)#switchport trunk encapsulation dot1q
sw-j1(config-if)#switchport trunk allowed vlan all
```

sw-j1(config-if)#speed 100
sw-j1(config-if)#duplex full
sw-j1(config-if)#no shutdown
sw-j1(config-if)# interface range f1/1 - 5
sw-j1(config-if-range)#switchport mode access
sw-j1(config-if-range)#switchport access vlan 2
sw-j1(config-if-range)#speed 100
sw-j1(config-if-range)#duplex full
sw-j1(config-if-range)#no shutdown
sw-j1(config-if)# interface range f1/6 - 10
sw-j1(config-if-range)#switchport mode access
sw-j1(config-if-range)#switchport access vlan 3
sw-j1(config-if-range)#speed 100
sw-j1(config-if-range)#duplex full
sw-j1(config-if-range)#no shutdown
sw-j1(config-if)#end
sw-j1#write

（3）交换机 sw-j2 的配置。
switch>enable
switch#config terminal
switch(config)# hostname sw-j1
sw-j2(config)#exit
sw-j2#vlan database
sw-j2(vlan)#vtp domain test
sw-j2(vlan)#vtp client
sw-j2(vlan)#vtp password abc
sw-j2(vlan)#exit
sw-j2#config terminal
sw-j2(config)# interface f1/0
sw-j2(config-if)#switchport mode trunk
sw-j2(config-if)#switchport trunk encapsulation dot1q
sw-j2(config-if)#switchport trunk allowed vlan all
sw-j2(config-if)#speed 100
sw-j2(config-if)#duplex full
sw-j2(config-if)#no shutdown
sw-j2(config-if)# interface range f1/1 - 5
sw-j2(config-if-range)#switchport mode access
sw-j2(config-if-range)#switchport access vlan 3
sw-j2(config-if-range)#speed 100
sw-j2(config-if-range)#duplex full
sw-j2(config-if-range)#no shutdown
sw-j2(config-if)# interface range f1/6 - 10
sw-j2(config-if-range)#switchport mode access
sw-j2(config-if-range)#switchport access vlan 4
sw-j2(config-if-range)#speed 100
sw-j2(config-if-range)#duplex full
sw-j2(config-if-range)#no shutdown

sw-j2(config-if)#end
sw-j2#write

## 七、实验测试和查看相关配置

（1）交换机 sw-h1 的 VTP 的配置相关信息。

```
sw-h1#show vtp status
VTP Version                    : 2      #VTP 协议版本号
Configuration Revision         : 2      #配置编号
Maximum VLANs supported locally: 256    #支持最大的 VLAN 数量
Number of existing VLANs       : 8      #已经存在的 VLAN 数量
VTP Operating Mode             : Server #当前交换机的模式为 Server
VTP Domain Name                : test   #当前交换机所在的域为 test
VTP Pruning Mode               : Enabled #启用了裁剪功能
VTP V2 Mode                    : Enabled #当前运行的版本为 2
VTP Traps Generation           : Disabled #向网络管理站发送 VTP 陷阱不可用
MD5 digest                     : 0xC3 0x8F 0xE9 0x3A 0xB9 0x5C 0x3D 0xCD  #启用 VTP 协议认证
Configuration last modified by 0.0.0.0 at 3-1-02 02:58:52  #最后修改 VTP 配置信息的时间
Local updater ID is 0.0.0.0 (no valid interface found)
```

（2）交换机 sw-h1 的 trunk 端口的配置相关信息。

Port：端口。
Mode：Trunk 模式的状态，on 表示运行模式为 Trunk 模式。
Encapsulation：端口封装的协议。
Status：运行的状态，trunking 表示正在运行 Trunk 模式。
Native vlan：本地 VLAN。

sw-h1#show interfaces f1/1 trunk

| Port  | Mode | Encapsulation | Status   | Native vlan |
|-------|------|---------------|----------|-------------|
| Fa1/1 | on   | 802.1q        | trunking | 1           |

| Port  | Vlans allowed on trunk | #允许那些 VLAN 信息通过此端口 |
|-------|------------------------|------------------------------|
| Fa1/1 | 1-3,5-1005             |                              |

| Port  | Vlans allowed and active in management domain | #在管理域中允许和激活的 VLAN |
|-------|-----------------------------------------------|------------------------------|
| Fa1/1 | 1-3                                           |                              |

| Port  | Vlans in spanning tree forwarding state and not pruned | #在生成树协议中处于转发状态的 VLAN，没有启用裁剪功能 |
|-------|--------------------------------------------------------|------------------------------------------------------|
| Fa1/1 | 1-3                                                    |                                                      |

（3）交换机 sw-j1 的 VTP 的配置相关信息。

```
sw-j1#show vtp status
VTP Version                    : 2
Configuration Revision         : 2
Maximum VLANs supported locally: 256
Number of existing VLANs       : 8
VTP Operating Mode             : Client
```

| | |
|---|---|
| VTP Domain Name | : test |
| VTP Pruning Mode | : Enabled |
| VTP V2 Mode | : Enabled |
| VTP Traps Generation | : Disabled |
| MD5 digest | : 0xC3 0x8F 0xE9 0x3A 0xB9 0x5C 0x3D 0xCD |

Configuration last modified by 0.0.0.0 at 3-1-02 02:58:52

（4）交换机 sw-j1 的 trunk 端口的配置相关信息。

sw-j1#show interfaces f1/0 trunk

| Port | Mode | Encapsulation | Status | Native vlan |
|---|---|---|---|---|
| Fa1/0 | on | 802.1q | trunking | 1 |

| Port | Vlans allowed on trunk |
|---|---|
| Fa1/0 | 1-3,5-1005 |

| Port | Vlans allowed and active in management domain |
|---|---|
| Fa1/0 | 1-3 |

| Port | Vlans in spanning tree forwarding state and not pruned |
|---|---|
| Fa1/0 | 1-3 |

（5）交换机 sw-j1 的 VLAN 的相关信息。

sw-j1#show vlan-switch brief

| VLAN | Name | Status | Ports |
|---|---|---|---|
| 1 | default | active | Fa1/11, Fa1/12, Fa1/13, Fa1/14 Fa1/15 |
| 2 | VLAN0002 | active | Fa1/1, Fa1/2, Fa1/3, Fa1/4 Fa1/5 |
| 3 | VLAN0003 | active | Fa1/6, Fa1/7, Fa1/8, Fa1/9 Fa1/10 |
| 4 | VLAN0004 | active | |
| 1002 | fddi-default | active | |
| 1003 | trcrf-default | active | |
| 1004 | fddinet-default | active | |
| 1005 | trbrf-default | active | |

（6）交换机 sw-j2 的 VTP 的配置相关信息。

sw-j2#show vtp status

| | |
|---|---|
| VTP Version | : 2 |
| Configuration Revision | : 2 |
| Maximum VLANs supported locally | : 256 |
| Number of existing VLANs | : 8 |
| VTP Operating Mode | : Client |
| VTP Domain Name | : test |
| VTP Pruning Mode | : Disabled |
| VTP V2 Mode | : Enabled |
| VTP Traps Generation | : Disabled |

MD5 digest                             : 0xC3 0x8F 0xE9 0x3A 0xB9 0x5C 0x3D 0xCD
Configuration last modified by 0.0.0.0 at 3-1-02 02:58:52

（7）交换机 sw-j2 的 trunk 端口的配置相关信息。

sw-j2#show interfaces f1/0 trunk

| Port  | Mode | Encapsulation | Status   | Native vlan |
|-------|------|---------------|----------|-------------|
| Fa1/0 | on   | 802.1q        | trunking | 1           |

| Port  | Vlans allowed on trunk |
|-------|------------------------|
| Fa1/0 | 1,3-1005               |

| Port  | Vlans allowed and active in management domain |
|-------|-----------------------------------------------|
| Fa1/0 | 1,3-4                                         |

| Port  | Vlans in spanning tree forwarding state and not pruned |
|-------|--------------------------------------------------------|
| Fa1/0 | 1,3-4                                                  |

（8）交换机 sw-j2 的 VLAN 的相关信息。

sw-j2#show vlan-switch brief

| VLAN | Name           | Status | Ports                              |
|------|----------------|--------|------------------------------------|
| 1    | default        | active |                                    |
| 2    | VLAN0002       | active | Fa1/11, Fa1/12, Fa1/13, Fa1/14     |
|      |                |        | Fa1/15                             |
| 3    | VLAN0003       | active | Fa1/1, Fa1/2, Fa1/3, Fa1/4         |
|      |                |        | Fa1/5                              |
| 4    | VLAN0004       | active | Fa1/6, Fa1/7, Fa1/8, Fa1/9         |
|      |                |        | Fa1/10                             |
| 1002 | fddi-default   | active |                                    |
| 1003 | trcrf-default  | active |                                    |
| 1004 | fddinet-default| active |                                    |
| 1005 | trbrf-default  | active |                                    |

## 八、注意问题

为了保证 VTP 域的安全，VTP 域需要设置密码，同一个域中所有交换机必须设置成一样的密码。在 VTP 域中的交换机配置了同样的密码后，VTP 才能正常工作。而不知道密码或密码错误的交换机将无法获知 VLAN 的消息。

# 实验七　链路聚合的配置

把多个物理链接捆绑在一起形成一个逻辑链接，这个逻辑链接我们称之为 Port Channel。Port Channel 功能符合 IEEE 802.3ad 标准，它可以用于扩展链路带宽，提供更高的连接可靠性。

当 Port Channel 中的一条成员链路断开时，系统会将该成员链路的流量自动地分配到 Port Channel 中的其他有效成员链路上去。Port Channel 中一条成员链路收到的广播或者多播报文，将不会被转发到其他成员链路上。

## 一、实验目的

交换机中链路聚合的配置，增加链路的带宽与链路的冗余。

## 二、实验内容

（1）链路聚合的配置；
（2）配置链路聚合的负载均衡的算法。

## 三、工作原理

端口聚合也叫做以太通道（Ethernet Channel），主要用于交换机之间的连接。由于当两个交换机之间有多条冗余链路的时候，STP 会将其中的几条链路关闭，只保留一条，这样可以避免二层的环路产生。但是失去了路径冗余的优点，因为 STP 的链路切换会很慢，在 50 秒左右。使用以太通道的话，交换机会把一组物理端口联合起来，作为一个逻辑的通道，也就是 channel-group，这样交换机会认为这个逻辑通道为一个端口。

（1）端口聚合。

端口聚合是将多个端口聚合在一起，形成 1 个汇聚组，以实现出负荷在各成员端口中的分担，同时也提供了更高的连接可靠性。端口聚合可以分为手工汇聚、动态 LACP 汇聚和静态 LACP 汇聚。同一个汇聚组中端口的基本配置应该保持一致，即如果某端口为 trunk 端口，则其他端口也配置为 trunk 端口；如该端口的链路类型改为 access 端口，则其他端口的链路类型也改为 access 端口。

端口的基本配置主要包括 STP、QOS、VLAN、端口属性等相关配置。其中 STP 配置包括：端口的 STP 使能/关闭、与端口相连的链路属性（如点对点或非点对点）、STP 优先级、路径开销、报文发送速率限制、是否环路保护、是否根保护、是否为边缘端口。QOS 配置包括：流量限速、优先级标记、默认的 IEEE 802.1p 优先级、带宽保证、拥塞避免、流重定向、流量统计等。VLAN 配置包括：端口上允许通过的 VLAN、端口默认 VLAN ID。端口属性配置包括：端口的链路类型（如 trunk、hybrid、access 属性）、绑定侦测组配置。一台 s9500 系列路由交换机最多可以配置 920 个汇聚组，其中 1~31 为手工或者静态聚合组；32~64 为预留组；65~192 为 routed trunk；193~920 为动态聚合组。

（2）LACP 协议。

基于 IEEE 802.3ad 标准的 LACP（Link Aggregation Control Protocol，链路聚合控制协议）

是一种实现链路动态聚合与解聚合的协议。LACP 协议通过 LACPDU(Link Aggregation Control Protocol Data Unit，链路聚合控制协议数据单元）与对端交互信息。使用某端口的 LACP 协议后，该端口将通过发送 LACPDU 向对端通告自己的系统优先级、系统 MAC、端口优先级、端口号和操作 key。对端接收到这些信息后，将这些信息与其他端口所保存的信息比较，以选择能够聚合的端口，从而双方可以对端口加入或退出某个动态聚合组达成一致。操作 key 是在端口聚合时，LACP 协议根据端口的配置（即速率、双工、基本配置、管理 key）生成的一个配置组合。其中，动态聚合端口在使用 LACP 协议后，其管理 key 默认为零。静态聚合端口在使用 LACP 后，端口的管理 key 与聚合组 ID 相同。对于动态聚合组而言，同组成员一定有相同的操作 key，而手工和静态聚合组中，selected 的端口有相同的操作 key。

（3）技术优点。

1）带宽增加。带宽相当于组成组的端口的带宽总和。

2）增加冗余。只要组内不是所有的端口都 down 掉，两个交换机之间仍然可以继续通信。

3）负载均衡。可以在组内的端口上配置，使流量可以在这些端口上自动进行负载均衡。

## 四、实验需要掌握的命令

switch(config)#interface port-channel *number*
#创建一个端口号为 *number* 的 port-channel，或者进入端口号为 *number* 的 port-channel
switch(config-if) # channel-group *number* mode on
#将指定的端口划分给端口号为 *number* 的 port-channel
switch (config)#port-channel load-balance dst-ip {dst-mac | src-dst-ip | src-dst-mac | src-ip | src-mac}
#设置 port-channel 的负载均衡的算法
switch#show ethernetchannel load-balance
#查看 port-channel 的负载均衡的算法
switch#show ethernetchannel [ *number* ] port-channel
#查看 port-channel 的端口信息
switch#show ethernetchannel [ *number* ] summary
#查看 port-channel 端口与物理端口关联的相关信息
switch#show ethernetchannel [ *number* ] detail
#查看 port-channel 端口详细信息
switch#show interface port-channel *number* etherchannel
#查看 port-channel 端口号为 *number* 的端口与物理端口关联的相关信息

## 五、实验拓扑图与网络文件

（1）网络拓扑图。

网络拓扑图如图 7-1 所示。

（2）网络文件。

#port-channel.net START
autostart = false
ghostios = true

图 7-1 网络拓扑图

sparsemem = true

[localhost]

    [[3640]]
    image = ..\C3640-IK.BIN
    ram = 160

    [[1720]]
    image = ..\C1700-Y.BIN
    ram = 16

    [[router sw1]]
    model = 3640
    idlepc = 0x60577fb4
    slot1 = NM-16ESW
    f1/0 = pc1 f0
    f1/2 = sw2 f1/2
    f1/1 = sw2 f1/1

    [[router sw2]]
    model = 3640
    idlepc = 0x60577fb4
    slot1 = NM-16ESW
    f1/0 = pc2 f0

    [[router pc0]]
    model = 1720
    idlepc = 0x8014e4ec

    [[router pc1]]
    model = 1720
    idlepc = 0x8014e4ec
#port-channel.net END

## 六、实验配置步骤

（1）交换机 sw1 的配置。

```
switch>enable
switch #configure terminal
switch (config)#hostname sw1
sw1(config)#interface port-channel 1
```

sw1(config)#no shutdown
sw1(config-if)# interface range FastEthernet 1/1 - 2
sw1(config-if-range)# channel-group 1 mode on
sw1(config-if-range)#no shutdown
sw1(config)#port-channel load-balance dst-ip
sw1(config)#end
sw1#write

（2）交换机 sw2 的配置。

switch>enable
switch #configure terminal
switch (config)#hostname sw2
sw2(config)#interface port-channel 1
sw2(config)#no shutdown
sw2(config-if)# interface range FastEthernet 1/1 - 2
sw2(config-if-range)# channel-group 1 mode on
sw2(config-if-range)#no shutdown
sw2(config)#port-channel load-balance dst-ip
sw2(config)#end
sw2#write

（3）主机 pc0 的配置。

interface FastEthernet0
　ip address 192.168.1.1 255.255.255.0
　no ip directed-broadcast
　full-duplex

（4）主机 pc1 的配置。

interface FastEthernet0
　ip address 192.168.1.2 255.255.255.0
　no ip directed-broadcast
　full-duplex

## 七、实验测试和查看相关配置

（1）查看交换机 sw1 的 port-channel 1 端口的配置。

sw1#show interfaces port-channel 1
Port-channel1 is **up**, line protocol is **up**　　#端口已经启用，端口线路协议正常
　Hardware is EtherChannel, address is cc00.0fe8.f101 (bia cc00.0fe8.f101)
　MTU 1500 bytes, BW 200000 Kbit, DLY 1000 usec,
　　　reliability 255/255, txload 1/255, rxload 1/255
　Encapsulation ARPA, loopback not set
　Keepalive set (10 sec)
　Full-duplex, 100Mb/s
　Members in this channel: Fa1/1 Fa1/2
　ARP type: ARPA, ARP Timeout 04:00:00
　Last input 00:22:44, output never, output hang never
　Last clearing of "show interface" counters never
　Input queue: 0/75/0/0 (size/max/drops/flushes); Total output drops: 0
　Queueing strategy: fifo

```
     Output queue: 0/40 (size/max)
     5 minute input rate 0 bits/sec, 0 packets/sec
     5 minute output rate 0 bits/sec, 0 packets/sec
        0 packets input, 0 bytes, 0 no buffer
        Received 0 broadcasts, 0 runts, 0 giants, 0 throttles
        0 input errors, 0 CRC, 0 frame, 0 overrun, 0 ignored
        0 input packets with dribble condition detected
        0 packets output, 0 bytes, 0 underruns
        0 output errors, 0 collisions, 1 interface resets
        0 babbles, 0 late collision, 0 deferred
        0 lost carrier, 0 no carrier
        0 output buffer failures, 0 output buffers swapped out
sw1#show interfaces port-channel 1 etherchannel
```

Index：索引编号
Port：端口编号
EC state：当前端口的状态，on 为可用

```
Age of the Port-channel    = 00d:00h:30m:51s
Logical slot/port   = 5/0            Number of ports = 2
GC                  = 0x00010001     HotStandBy port = null
Port state          = Port-channel Ag-Inuse

Ports in the Port-channel:

Index    Port    EC state
------+------+------------
  0      Fa1/1    on
  1      Fa1/2    on


Time since last port bundled:     00d:00h:30m:50s    Fa1/2
```

（2）显示链路聚合的算法。

```
sw1#show etherchannel load-balance
  Po1 ---> Destination IP address    #Port Channel 1 使用的流量平衡算法为基于目的 IP 地址
```

（3）测试链路聚合的运行结果。

首先在 dynagen 窗口输入 capture sw1 f1/1 f1.cap 和 capture sw1 f1/2 f2.cap，如图 7-2 所示。

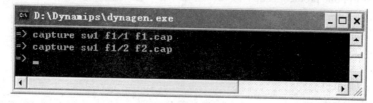

图 7-2　抓取数据包

此时在网络拓扑文件所在的目录下会出现两个文件：f1.cap 和 f2.cap；然后在 pc0 上 ping pc1；最后用 Wireshark 软件分别打开两个文件 f1.cap 和 f2.cap，可以看到如图 7-3 和图 7-4 所示的结果。

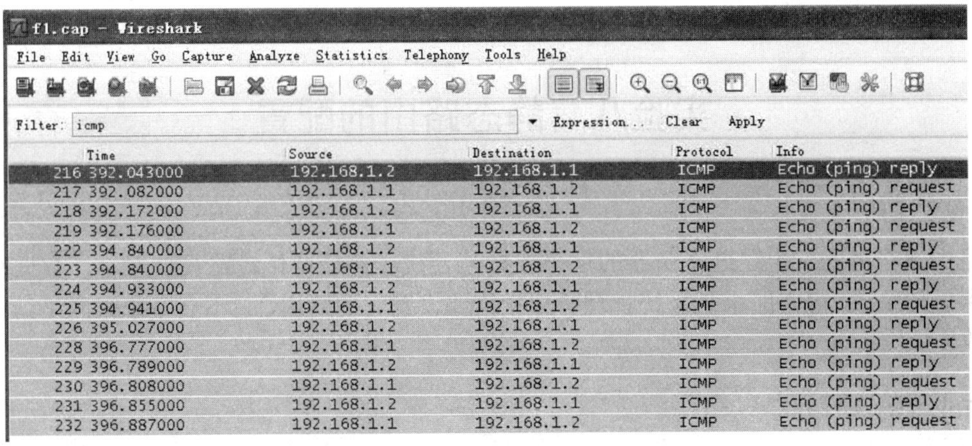

图 7-3　显示数据包 f1.cap

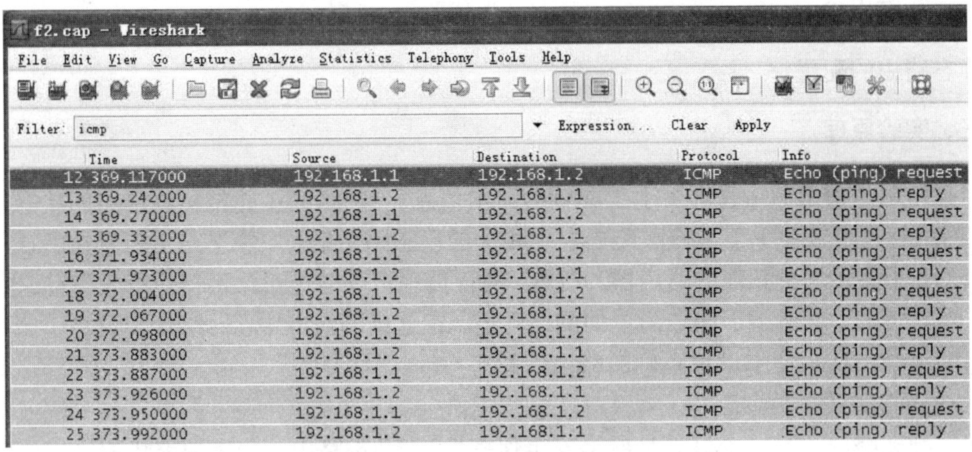

图 7-4　显示数据包 f2.cap

从图 7-3 和图 7-4 中可以看出，数据包有一半通过交换机 sw1 的 f1/1 进行传输，另一半通过交换机 sw1 的 f1/2 进行传输。

## 八、注意问题

要查看链路聚合的算法、port-channel 端口的状态和线路协议必须是 up。

# 实验八  静态路由的配置

路由的作用就是为经过路由器的每个数据帧寻找一条最佳传输路径,并将该数据有效地传送到目的站点。静态路由是系统管理员根据需要设置好的路由,它不会随未来网络结构的改变而改变。

## 一、实验目的

理解路由的作用,掌握静态路由的配置。

## 二、实验内容

静态路由的配置。

## 三、实验原理

静态路由是指由网络管理员手工配置的路由信息。当网络的拓扑结构或链路的状态发生变化时,网络管理员需要手工去修改路由表中相关的静态路由信息。静态路由信息在默认情况下是私有的,不会传递给其他的路由器。当然,网管员也可以通过对路由器进行设置使之成为共享的。静态路由一般适用于比较简单的网络环境,在这样的环境中,网络管理员易于清楚地了解网络的拓扑结构,便于设置正确的路由信息。

(1)静态路由的优点。

使用静态路由的另一个好处是网络安全保密性高。动态路由因为需要路由器之间频繁地交换各自的路由表,而对路由表的分析可以揭示网络的拓扑结构和网络地址等信息。因此,网络出于安全方面的考虑也可以采用静态路由。

(2)静态路由的缺点。

大型和复杂的网络环境通常不宜采用静态路由。一方面,网络管理员难以全面地了解整个网络的拓扑结构;另一方面,当网络的拓扑结构和链路状态发生变化时,路由器中的静态路由信息需要大范围地调整,这一工作的难度和复杂程度非常高。

(3)路由匹配原则。

1)最长匹配。

路由掩码最长匹配原则是指 IP 网络中,当路由表中有多条条目可以匹配目的 IP 时,一般就采用掩码最长的一条作为匹配项并确定下一跳。

例如,考虑下面这个 IPv4 的路由表:

192.168.19.16/28 e0

192.168.0.0/16 s0

在要查找地址 192.168.19.19 时,不难发现上述两条都"匹配",即这两条都包含要查找的目的地址。此时就应该根据最长掩码匹配原则,选择第一条进行匹配(更明确),所以数据包将通过 e0 发送出去。

2）最小管理距离优先。

在相同匹配长度的情况下，按照路由的管理距离进行匹配：管理距离越小，路由越优先。例如 S 10.1.1.1/8 为静态路由，R 10.1.1.1/8 为 RIP 产生的动态路由，静态路由的默认管理距离值为 1，而 RIP 默认管理距离为 120，因而选 S 10.1.1.1/8。

3）度量值最小优先。

当匹配长度、管理距离都相同时，比较路由的度量值（Metric）或称代价，度量值越小越优先。例如：S 10.1.1.1/8[1/20]，其度量值为 20；S 10.1.1.1/8 [1/40]，其度量值为 40，因而选 S 10.1.1.1/8[1/20]。

### 四、实验需要掌握的命令

switch (config)#no ip routing
#关闭路由功能
switch (config)#ip default-gateway *IP-address*
#设置默认网关，其中 *IP-address* 为网关地址
switch (config)#ip route *Destination-IP-address subnet-mask* {*IP—address|interface-number*}
#设置路由，其中 *Destination-IP-address* 为目的网段地址，*subnet-mask* 为目的网段的子网掩码，*IP-address* 为下一跳的 IP 地址，*interface-number* 为下一跳的端口
switch (config)no ip route *Destination-IP-address subnet-mask forward-address*
#删除一条路由

### 五、实验拓扑图和网络文件

（1）网络结构图。

网络结构图如图 8-1 所示。

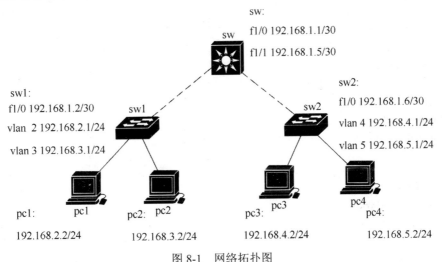

图 8-1　网络拓扑图

（2）网络文件。
# test.net 文件 START
autostart = false
ghostios = true

sparsemem = true

[localhost]

    [[3660]]
    image = ..\C3660-IS.BIN
    ram = 160

    [[1720]]
    image = ..\C1700-Y.BIN
    ram = 16

    [[router sw]]
    model = 3660
    idlepc = 0x60577fb4
    slot1 = NM-16ESW
    f1/0 = sw1 f1/0
    f1/1 = sw2 f1/0

    [[router sw1]]
    model = 3660
    idlepc = 0x60577fb4
    slot1 = NM-16ESW
    f1/1 = pc1 f0
    f1/2 = pc2 f0

    [[router sw2]]
    model = 3660
    idlepc = 0x60577fb4
    slot1 = NM-16ESW
    f1/1 = pc3 f0
    f1/2 = pc4 f0

    [[router pc1]]
    model = 1720
    idlepc = 0x8014e4ec

    [[router pc2]]
    model = 1720
    idlepc = 0x8014e4ec

    [[router pc3]]
    model = 1720
    idlepc = 0x8014e4ec

    [[router pc4]]
    model = 1720
    idlepc = 0x8014e4ec

# test.net 文件 END

## 六、实验配置步骤

（1）交换机 sw 的配置。

switch>enable
switch #configure terminal
switch (config)#hostname sw
sw(config)#interface FastEthernet 1/0
sw(config-if)#no switchport
sw(config-if)#ip address 192.168.1.1 255.255.255.252
sw(config-if)#speed 100
sw(config-if)#duplex full
sw(config-if)#no shutdown
sw(config-if)#exit
sw(config)#interface FastEthernet 1/1
sw(config-if)#no switchport
sw(config-if)#ip address 192.168.1.5 255.255.255.252
sw(config-if)#speed 100
sw(config-if)#duplex full
sw(config-if)#no shutdown
sw(config-if)#exit
sw(config)#ip routing
sw(config)#ip route 192.168.2.0 255.255.255.0 192.168.1.2
sw(config)#ip route 192.168.3.0 255.255.255.0 192.168.1.2

（上面的两条路由可以用下面一条聚合后的路由来代替）

**sw(config)#ip route 192.168.2.0 255.255.254.0 192.168.1.2）**

sw(config)#ip route 192.168.4.0 255.255.255.0 192.168.1.6
sw(config)#ip route 192.168.5.0 255.255.255.0 192.168.1.6

（上面的两条路由可以用下面一条聚合后的路由来代替）

**sw(config)#ip route 192.168.4.0 255.255.254.0 192.168.1.6）**

sw(config)#exit
sw#write

（2）交换机 sw1 的配置。

switch>enable
switch #configure terminal
switch (config)#hostname sw1
sw1(config)#interface FastEthernet 1/0
sw1(config-if)#no switchport
sw1(config-if)#ip address 192.168.1.2 255.255.255.252
sw1(config-if)#speed 100
sw1(config-if)#duplex full
sw1(config-if)#no shutdown
sw1(config-if)#exit
sw1(config)#ip routing
sw1(config)#ip route 192.168.4.0 255.255.255.0 192.168.1.1
sw1(config)#ip route 192.168.5.0 255.255.255.0 192.168.1.1

（上面的两条路由可以用下面一条聚合后的路由来代替）

**sw1(config)#ip route 192.168.4.0 255.255.254.0 192.168.1.1)**
sw1(config)#ip route 192.168.1.4 255.255.255.252 192.168.1.1
sw1(config)#exit
sw1#vlan database
sw1(vlan)#vlan 2
sw1(vlan)#vlan 3
sw1(vlan)#exit
sw1#configure terminal
sw1(config)#interface FastEthernet 1/1
sw1(config-if)#switchport
sw1(config-if)#switchport mode access
sw1(config-if)#switchport access vlan 2
sw1(config-if)#no shutdown
sw1(config-if)#interface FastEthernet 1/2
sw1(config-if)#switchport
sw1(config-if)#switchport mode access
sw1(config-if)#switchport access vlan 3
sw1(config-if)#no shutdown
sw1(config-if)#exit
sw1(config)#interface vlan 2
sw1(config-if)#ip address 192.168.2.1 255.255.255.0
sw1(config-if)#no shutdown
sw1(config-if)#exit
sw1(config)# interface vlan 3
sw1(config-if)#ip address 192.168.3.1 255.255.255.0
sw1(config-if)#no shutdown
sw1(config-if)#end
sw1#write

（3）交换机 sw2 的配置。
switch>enable
switch #configure terminal
switch (config)#hostname sw2
sw2(config)#interface FastEthernet 1/0
sw2(config-if)#no switchport
sw2(config-if)#ip address 192.168.1.6 255.255.255.252
sw2(config-if)#speed 100
sw2(config-if)#duplex full
sw2(config-if)#no shutdown
sw2(config-if)#exit
sw2(config)#ip routing
sw2(config)#ip route 192.168.2.0 255.255.255.0 192.168.1.5
sw2(config)#ip route 192.168.3.0 255.255.255.0 192.168.1.5
（上面的两条路由可以用下面一条聚合后的路由来代替）
**sw2(config)#ip route 192.168.2.0 255.255.254.0 192.168.1.5)**
sw2(config)#ip route 192.168.1.0 255.255.255.252 192.168.1.5
sw2(config)#exit
sw2#vlan database
sw2(vlan)#vlan 4

sw2(vlan)#vlan 4
sw2(vlan)#exit
sw2#configure terminal
sw2(config)#interface FastEthernet 1/1
sw2(config-if)#switchport
sw2(config-if)#switchport mode access
sw2(config-if)#switchport access vlan 4
sw2(config-if)#no shutdown
sw2(config-if)#interface FastEthernet 1/2
sw2(config-if)#switchport
sw2(config-if)#switchport mode access
sw2(config-if)#switchport access vlan 4
sw2(config-if)#no shutdown
sw2(config-if)#exit
sw2(config)#interface vlan 4
sw2(config-if)#ip address 192.168.4.1 255.255.255.0
sw2(config-if)#no shutdown
sw2(config-if)#exit
sw2(config)# interface vlan 5
sw2(config-if)#ip address 192.168.5.1 255.255.255.0
sw2(config-if)#no shutdown
sw2(config-if)#end
sw2#write

（4）主机 pc1 的配置。

switch>enable
switch #configure terminal
switch (config)#hostname pc1
pc1(config)#interface FastEthernet 0
pc1(config-if)# ip address 192.168.2.2 255.255.255.0
pc1(config-if)# duplex full
pc1(config-if)#exit
pc1(config)#no ip routing
pc1(config)#ip default-gateway 192.168.2.1
pc1(config)#exit
pc1#write

（5）主机 pc2 的配置。

switch>enable
switch #configure terminal
switch (config)#hostname pc2
pc2(config)#interface FastEthernet 0
pc2(config-if)# ip address 192.168.3.2 255.255.255.0
pc2(config-if)# duplex full
pc2(config-if)#exit
pc2(config)#no ip routing
pc2(config)#ip default-gateway 192.168.3.1
pc2(config)#exit
pc2#write

（6）主机 pc3 的配置。

switch>enable
switch #configure terminal
switch (config)#hostname pc3
pc3(config)#interface FastEthernet 0
pc3(config-if)# ip address 192.168.4.2 255.255.255.0
pc3(config-if)# duplex full
pc3(config-if)#exit
pc3(config)#no ip routing
pc3(config)#ip default-gateway 192.168.4.1
pc3(config)#exit
pc3#write

（7）主机 pc4 的配置。

switch>enable
switch #configure terminal
switch (config)#hostname pc4
pc4(config)#interface FastEthernet 0
pc4(config-if)# ip address 192.168.5.2 255.255.255.0
pc4(config-if)# duplex full
pc4(config-if)#exit
pc4(config)#no ip routing
pc4(config)#ip default-gateway 192.168.5.1
pc4(config)#exit
pc4#write

## 七、实验测试和查看相关配置

（1）查看交换机 sw 的端口状态。

sw#show interfaces
……
FastEthernet1/0 is up, line protocol is up
　　Hardware is FastEthernet, address is cc00.0d8c.f100 (bia cc00.0d8c.f100)
　　Internet address is 192.168.1.1/30
　　MTU 1500 bytes, BW 100000 Kbit, DLY 100 usec,
　　　　reliability 255/255, txload 1/255, rxload 1/255
　　Encapsulation ARPA, loopback not set
　　Keepalive set (10 sec)
　　Full-duplex, 100Mb/s
　　……
FastEthernet1/1 is up, line protocol is up
　　Hardware is FastEthernet, address is cc00.0d8c.f101 (bia cc00.0d8c.f101)
　　Internet address is 192.168.1.5/30
　　MTU 1500 bytes, BW 100000 Kbit, DLY 100 usec,
　　　　reliability 255/255, txload 1/255, rxload 1/255
　　Encapsulation ARPA, loopback not set
　　Keepalive set (10 sec)
　　Full-duplex, 100Mb/s
　　……

## 实验八 静态路由的配置

（2）查看交换机 sw 的路由信息。

S　　192.168.4.0/24 [1/0] via 192.168.1.6：路由中 S 的意思是此条路由是静态路由；192.168.4.0/24，其中 192.168.4.0 是目的网段地址，/24 的意思是目的网段子网掩码为 24 位；[1/0]的意思是管理距离为 1，度量值为 0；via 192.168.1.6 的意思是下一跳的地址为 192.168.1.6。本条路由的意思是要到达目的网段 192.168.4.0/24 子网需要通过 192.168.1.6 地址进行转发。

192.168.1.0/30 is subnetted, 2 subnets 的意思是在 192.168.1.0 网段有两个子网，子网掩码为 30 位。

C　　192.168.1.0 is directly connected, FastEthernet1/0：路由中 C 代表是此条路由是直连路由；本条路由的意思是 192.168.1.0/30 子网和路由器直接相连，直接相连的接口是 FastEthernet1/0。

```
sw#sho ip route
Codes: C - connected, S - static, R - RIP, M - mobile, B - BGP
       D - EIGRP, EX - EIGRP external, O - OSPF, IA - OSPF inter area
       N1 - OSPF NSSA external type 1, N2 - OSPF NSSA external type 2
       E1 - OSPF external type 1, E2 - OSPF external type 2
       i - IS-IS, su - IS-IS summary, L1 - IS-IS level-1, L2 - IS-IS level-2
       ia - IS-IS inter area, * - candidate default, U - per-user static route
       o - ODR, P - periodic downloaded static route

Gateway of last resort is not set

S       192.168.4.0/24 [1/0] via 192.168.1.6
S       192.168.5.0/24 [1/0] via 192.168.1.6
        192.168.1.0/30 is subnetted, 2 subnets
C          192.168.1.0 is directly connected, FastEthernet1/0
C          192.168.1.4 is directly connected, FastEthernet1/1
S       192.168.2.0/24 [1/0] via 192.168.1.2
S       192.168.3.0/24 [1/0] via 192.168.1.2
```

（3）查看交换机 sw1 的端口状态。

```
sw1#show interfaces
……
FastEthernet1/0 is up, line protocol is up
    Hardware is Fast Ethernet, address is cc01.0d8c.f100 (bia cc01.0d8c.f100)
    Internet address is 192.168.1.2/30
    MTU 1500 bytes, BW 100000 Kbit, DLY 100 usec,
        reliability 255/255, txload 1/255, rxload 1/255
    Encapsulation ARPA, loopback not set
    Keepalive set (10 sec)
    Full-duplex, 100Mb/s
    ……
FastEthernet1/1 is up, line protocol is up
    ……
FastEthernet1/2 is up, line protocol is up
    ……
Vlan2 is up, line protocol is up
    Hardware is EtherSVI, address is cc01.0d8c.0000 (bia cc01.0d8c.0000)
    Internet address is 192.168.2.1/24
    ……
Vlan3 is up, line protocol is up
```

　　　　Hardware is EtherSVI, address is cc01.0d8c.0000 (bia cc01.0d8c.0000)
　　　　Internet address is 192.168.3.1/24
　　……

（4）查看交换机 sw1 的路由信息。

S　　192.168.4.0/23 [1/0] via 192.168.1.1：路由中 S 的意思是此条路由是静态路由；192.168.4.0/24，其中 192.168.4.0 是目的网段地址，/23 的意思是目的网段子网掩码为 23 位；[1/0]的意思是管理距离为 1，度量值为 0；via 192.168.1.1 的意思是下一跳的地址为 192.168.1.1。本条路由的意思是要到达目的网段 192.168.4.0/23 子网需要通过 192.168.1.1 地址进行转发。

192.168.1.0/30 is subnetted, 2 subnets 的意思是在 192.168.1.0 网段有两个子网，子网掩码为 30 位。

S　　192.168.1.4 [1/0] via 192.168.1.1：路由中 S 代表是此条路由是静态路由；本条路由的意思是要到达目的网段 192.168.1.4/30 子网需要通过 192.168.1.1 地址进行转发。

```
sw1#sho ip route
Codes: C - connected, S - static, R - RIP, M - mobile, B - BGP
       D - EIGRP, EX - EIGRP external, O - OSPF, IA - OSPF inter area
       N1 - OSPF NSSA external type 1, N2 - OSPF NSSA external type 2
       E1 - OSPF external type 1, E2 - OSPF external type 2
       i - IS-IS, su - IS-IS summary, L1 - IS-IS level-1, L2 - IS-IS level-2
       ia - IS-IS inter area, * - candidate default, U - per-user static route
       o - ODR, P - periodic downloaded static route

Gateway of last resort is not set

     192.168.1.0/30 is subnetted, 2 subnets
C       192.168.1.0 is directly connected, FastEthernet1/0
S       192.168.1.4 [1/0] via 192.168.1.1
C    192.168.2.0/24 is directly connected, Vlan2
C    192.168.3.0/24 is directly connected, Vlan3
S    192.168.4.0/23 [1/0] via 192.168.1.1
```

（5）查看交换机 sw2 的端口状态。

```
sw2#show interfaces
……
FastEthernet1/0 is up, line protocol is up
   Hardware is Fast Ethernet, address is cc02.0d8c.f100 (bia cc02.0d8c.f100)
   Internet address is 192.168.1.6/30
   MTU 1500 bytes, BW 100000 Kbit, DLY 100 usec,
      reliability 255/255, txload 1/255, rxload 1/255
   Encapsulation ARPA, loopback not set
   Keepalive set (10 sec)
   Full-duplex, 100Mb/s
……
FastEthernet1/1 is up, line protocol is up
……
FastEthernet1/2 is up, line protocol is up
……
Vlan4 is up, line protocol is up
   Hardware is EtherSVI, address is cc02.0d8c.0000 (bia cc02.0d8c.0000)
   Internet address is 192.168.4.1/24
```

……
Vlan5 is up, line protocol is up
　Hardware is EtherSVI, address is cc02.0d8c.0000 (bia cc02.0d8c.0000)
　Internet address is 192.168.5.1/24
……

（6）查看交换机 sw2 的路由信息。
S　　192.168.2.0/23 [1/0] via 192.168.1.5：路由中 S 的意思是此条路由是静态路由；192.168.2.0/24，其中 192.168.2.0 是目的网段地址，/23 的意思是目的网段子网掩码为 23 位；[1/0]的意思是管理距离为 1，度量值为 0；via 192.168.1.5 的意思是下一跳的地址为 192.168.1.5。本条路由的意思是要到达目的网段 192.168.2.0/23 子网需要通过 192.168.1.5 地址进行转发。
192.168.1.0/30 is subnetted, 2 subnets 的意思是在 192.168.1.0 网段有两个子网，子网掩码为 30 位。
C　　192.168.1.4 is directly connected, FastEthernet1/0：路由中 C 代表是此条路由是直连路由；本条路由的意思是 192.168.1.4/30 子网和路由器直接相连，直接相连的接口是 FastEthernet1/0。
sw2#show ip route
Codes: C - connected, S - static, R - RIP, M - mobile, B - BGP
　　　　D - EIGRP, EX - EIGRP external, O - OSPF, IA - OSPF inter area
　　　　N1 - OSPF NSSA external type 1, N2 - OSPF NSSA external type 2
　　　　E1 - OSPF external type 1, E2 - OSPF external type 2
　　　　i - IS-IS, su - IS-IS summary, L1 - IS-IS level-1, L2 - IS-IS level-2
　　　　ia - IS-IS inter area, * - candidate default, U - per-user static route
　　　　o - ODR, P - periodic downloaded static route

Gateway of last resort is not set

C　　192.168.4.0/24 is directly connected, Vlan4
C　　192.168.5.0/24 is directly connected, Vlan5
　　　192.168.1.0/30 is subnetted, 2 subnets
S　　　192.168.1.0 [1/0] via 192.168.1.5
C　　　192.168.1.4 is directly connected, FastEthernet1/0
S　　　192.168.2.0/23 [1/0] via 192.168.1.5

## 八、注意问题

（1）首先要保证每一对相连的端口能够 ping 通对方。
（2）在每台主机或交换机上保证每一个 IP 数据都有出路。

# 实验九 NAT 的配置

NAT（Network Address Translation，网络地址转换）是将 IP 数据报头中的源 IP 地址转换为另一个源 IP 地址，目的 IP 地址转换为另一个目的 IP 地址或者将源 IP 地址和目的 IP 地址同时进行转换的过程。在实际应用中，NAT 主要用于实现私有网络访问公共网络的功能。通过使用少量的公有 IP 地址代表较多的私有 IP 地址的方式，减缓可用 IP 地址空间的枯竭。

## 一、实验目的

（1）掌握 NAT 的作用和工作原理；
（2）掌握 NAT 的配置和排错。

## 二、实验内容

（1）掌握 NAT 的作用和工作原理；
（2）掌握 NAT 的配置和排错。

## 三、实验原理

NAT（Network Address Translation，网络地址转换）是一个 IETF（Internet Engineering Task Force，Internet 工程任务组）标准，允许一个整体机构以一个公用 IP（Internet Protocol）地址出现在 Internet 上。

简单地说，NAT 就是在局域网内部网络中使用内部地址，而当内部节点要与外部网络进行通信时，就在出口处将内部地址替换成公用地址，从而在外部公网（Internet）上正常使用，NAT 可以使多台计算机共享 Internet 连接，这一功能很好地解决了公共 IP 地址紧缺的问题。通过这种方法可以只申请一个合法 IP 地址，就把整个局域网中的计算机接入 Internet 中。这时，NAT 屏蔽了内部网络，所有内部网计算机对于公共网络来说是不可见的，而内部网计算机用户通常不会意识到 NAT 的存在。这里提到的内部地址是指在内部网络中分配给节点的私有 IP 地址，这个地址只能在内部网络中使用，不能被路由。虽然内部地址可以随机挑选，但是通常使用的是以下地址：10.0.0.0～10.255.255.255、172.16.0.0～172.31.255.255、192.168.0.0～192.168.255.255。NAT 将这些无法在互联网上使用的保留 IP 地址翻译成可以在互联网上使用的合法 IP 地址。而全局地址是指合法的 IP 地址，它是由 NIC（网络信息中心）或者 ISP（网络服务提供商）分配的地址，对外代表一个或多个内部局部地址，是全球统一的可寻址的地址。

NAT 有三种类型：静态 NAT、动态地址 NAT、网络地址端口转换 NAPT。

其中静态 NAT 是设置起来最为简单和最容易实现的一种，内部网络中的每个主机都被永久映射成外部网络中的某个合法的地址。

动态地址 NAT 只是转换 IP 地址，它为每一个内部的 IP 地址分配一个临时的外部 IP 地址，主要应用于拨号，对于频繁的远程连接也可以采用动态 NAT。当远程用户连接上之后，动态地址 NAT 就会分配给他一个 IP 地址，用户断开时，这个 IP 地址就会被释放而留待以后使用。

网络地址端口转换 NAPT（Network Address Port Translation）是人们比较熟悉的一种转换

方式。NAPT 普遍应用于接入设备中，它可以将中小型的网络隐藏在一个合法的 IP 地址后面。NAPT 与动态地址 NAT 不同，它将内部连接映射到外部网络中的一个单独的 IP 地址上，同时在该地址上加上一个由 NAT 设备选定的 TCP 端口号。

在 NAT 实验中需要理解以下术语：

内部局部地址（Inside Local）：在内部网络中分配给主机的私有 IP 地址；

内部全局地址（Inside Global）：一个合法的 IP 地址，它对外代表一个或多个内部局部 IP 地址；

外部局部地址（Outside Local）：外部主机映射到内部网络中的 IP 地址。

外部全局地址（Outside Global）：由其所有者给外部网络上的主机分配的 IP 地址；

NAT 的优点：

（1）对于那些家庭用户或者小型的商业机构来说，使用 NAT 可以更便宜、更有效率地接入 Internet；

（2）使用 NAT 可以缓解目前全球 IP 地址不足的问题；

（3）在很多情况下，NAT 能够满足安全性的需要；

（4）使用 NAT 可以方便网络的管理，并大大提高了网络的适应性。

NAT 的缺点：

（1）NAT 会增加延迟，因为要转换每个数据包包头的 IP 地址，自然要增加延迟；

（2）NAT 会使某些要使用内嵌地址的应用不能正常工作。

### 四、实验需要掌握的命令

switch (config-if)#ip nat inside
#设置接口为内部接口

switch (config-if)#ip nat outside
#设置接口为外内部接口

switch (config)#ip nat inside source static *local-ip global-ip*
#在内部局部地址使用静态地址转换时，将内部局部地址 *local-ip* 转换为内部全局地址 *global-ip*

switch (config)#access-list *access-list-number* {permit|deny} *local-ip subnet-wild*
#为内部网络定义一个标准的 IP 访问控制列表，其中 *access-list-number* 为访问列表编号，*local-ip* 为源地址，*subnet-wild* 为源地址网络的子网掩码的反码

switch (config)#ip nat pool *pool-name start-ip end-ip* { netmask *netmask* | prefix-length *number*} [type rotary]
#为内部网络定义一个 NAT 地址池

switch (config)#ip nat inside source list *access-list-number* pool *pool-name* [overload]
#定义访问控制列表与 NAT 内部全局地址池之间的映射

switch (config)#ip nat outside source list *access-list-number* pool *pool-name* [overload]
定义访问控制列表与 NAT 外部局部地址池之间的映射

switch (config)#ip nat inside destination list *access-list-number* pool *pool-name*
#定义访问控制列表与终端 NAT 地址池之间的映射

switch #show ip nat translations

#显示当前存在的 NAT 转换信息
switch #show ip nat statistics
#查看 NAT 的统计信息
switch #show ip nat translations verbose
#显示当前存在的 NAT 转换的详细信息
switch #debug ip nat
#跟踪 NAT 操作,显示出每个被转换的数据包
switch #clear ip nat translations *
#删除 NAT 映射表中的所有内容

### 五、实验拓扑图和网络文件

(1) 网络拓扑图。

网络拓扑图如图 9-1 所示。

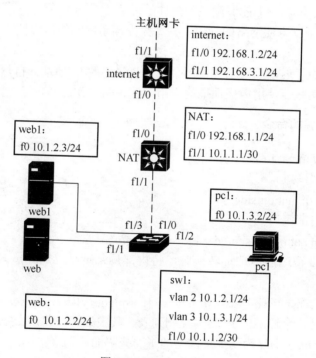

图 9-1 网络拓扑图

(2) 网络文件。
#nat.net 开始
autostart = false
ghostios = true
sparsemem = true


[localhost]

　　[[3660]]
　　image = ..\C3660-IS.BIN

ram = 160

[[1720]]
image = ..\C1700-Y.BIN
ram = 16

[[router internet]]
model = 3660
idlepc = 0x60577fb4
slot1 = NM-16ESW
f1/0 = NAT f1/0
f1/1 = NIO_gen_eth:\Device\NPF_{86530D2E-DF0E-4BC7-BAC3-8B956007903F}    #将 86530D2E-DF0E-4BC7-BAC3-8B956007903F 值修改为主机网卡的值，通过运行桌面上 Network device list 快捷方式得到

[[router NAT]]
model = 3660
idlepc = 0x60577fb4
slot1 = NM-16ESW
f1/1 = sw1 f1/0

[[router sw1]]
model = 3660
idlepc = 0x60577fb4
slot1 = NM-16ESW
f1/1 = web f0
f1/2 = pc1 f0
f1/3 = web1 f0

[[router pc1]]
model = 1720
idlepc = 0x8014e4ec

[[router web]]
model = 1720
idlepc = 0x8014e4ec

[[router web1]]
model = 1720
idlepc = 0x8014e4ec
#nat.net 结束

## 六、实验配置步骤

（1）基本配置。

1）Internet 交换机的配置。
switch>enable
switch #configure terminal

```
switch (config)#hostname internet
internet (config)#interface FastEthernet 1/0
internet (config-if)#no switchport
internet (config-if)#ip address 192.168.1.2 255.255.255.0
internet (config-if)#speed 100
internet (config-if)#duplex full
internet (config-if)#no shutdown
internet (config-if)#interface FastEthernet 1/1
internet (config-if)#no switchport
internet (config-if)#ip address 192.168.3.1 255.255.255.0
internet (config-if)#speed 100
internet (config-if)#duplex full
internet (config-if)#no shutdown
internet (config-if)#end
internet #write
```

2)NAT 三层交换机的配置。

```
switch>enable
switch #configure terminal
switch (config)#hostname NAT
NAT (config)#interface FastEthernet 1/0
NAT (config-if)#no switchport
NAT (config-if)#ip address 192.168.1.1 255.255.255.0
NAT (config-if)#speed 100
NAT (config-if)#duplex full
NAT (config-if)#ip nat outside
NAT (config-if)#no shutdown
NAT (config)#interface FastEthernet 1/1
NAT (config-if)#no switchport
NAT (config-if)#ip address 10.1.1.1 255.255.255.252
NAT (config-if)#speed 100
NAT (config-if)#duplex full
NAT (config-if)#ip nat inside
NAT (config-if)#no shutdown
NAT (config-if)#exit
NAT (config)#ip route 0.0.0.0 0.0.0.0 192.168.1.2
NAT (config)# ip route 10.1.2.0 255.255.254.0 10.1.1.2
NAT (config)#exit
NAT #write
```

3)sw1 交换机的配置。

```
switch>enable
switch #vlan database
switch (vlan) #vlan 2
switch (vlan) #vlan 3
switch (vlan) #exit
switch #configure terminal
switch (config)#hostname sw1
sw1 (config)#interface FastEthernet 1/0
```

sw1 (config-if)#no switchport
sw1 (config-if)#ip address 10.1.1.2 255.255.255.252
sw1 (config-if)#speed 100
sw1 (config-if)#duplex full
sw1 (config-if)#no shutdown
sw1 (config)#interface FastEthernet 1/1
sw1 (config-if)#switchport
sw1 (config-if)# switchport mode access
sw1 (config-if)# switchport access vlan 2
sw1 (config-if)#speed 100
sw1 (config-if)#duplex full
sw1 (config-if)#no shutdown
sw1 (config)#interface FastEthernet 1/2
sw1 (config-if)#switchport
sw1 (config-if)# switchport mode access
sw1 (config-if)# switchport access vlan 3
sw1 (config-if)#speed 100
sw1 (config-if)#duplex full
sw1 (config-if)#no shutdown
sw1 (config)#interface FastEthernet 1/23
sw1 (config-if)#switchport
sw1 (config-if)# switchport mode access
sw1 (config-if)# switchport access vlan 3
sw1 (config-if)#speed 100
sw1 (config-if)#duplex full
sw1 (config-if)#no shutdown
sw1 (config)#interface vlan 2
sw1 (config-if)# ip address 10.1.2.1 255.255.255.0
sw1 (config-if)#no shutdown
sw1 (config)#interface vlan 3
sw1 (config-if)# ip address 10.1.3.1 255.255.255.0
sw1 (config-if)#no shutdown
sw1 (config-if)#exit
sw1 (config)# ip route 0.0.0.0 0.0.0.0 10.1.1.1
sw1 (config)#exit
sw1 #write

4）web 服务器的配置。

switch>enable
switch#configure terminal
switch(config)#hostname web
web(config)#interface FastEthernet0
web(config-if)#ip address 10.1.2.2 255.255.255.0
web(config-if)#duplex full
web(config-if)#no shutdown
web(config-if)#exit
web(config)#no ip routing
web(config)#ip default-gateway 10.1.2.1

```
web(config)#ip http server
web(config)#exit
web#write
```

5）web1 服务器的配置。

```
switch>enable
switch #configure terminal
switch (config)#hostname web
web1(config)#interface FastEthernet0
web1(config-if)#ip address 10.1.2.3 255.255.255.0
web1(config-if)#duplex full
web1(config-if)#no shutdown
web1(config-if)#exit
web1(config)#no ip routing
web1(config)#ip default-gateway 10.1.2.1
web1(config)#ip http server
web1(config)#exit
web#write
```

6）pc1 主机的配置。

```
switch>enable
switch #configure terminal
switch (config)#hostname pc1
pc1 (config)#interface FastEthernet0
pc1 (config-if)#ip address 10.1.3.2 255.255.255.0
pc1 (config-if)#duplex full
pc1 (config-if)#no shutdown
pc1 (config-if)#exit
pc1 (config)#no ip routing
pc1 (config)#ip default-gateway 10.1.3.1
pc1 (config)#exit
pc1#write
```

（2）静态 NAT 的配置。

NAT 三层交换机的配置：

```
NAT>enable
NAT #configure terminal
NAT(config)#ip nat inside source static tcp 10.1.2.2 80 192.168.1.100 80
NAT(config)#exit
NAT#write
```

（3）动态 NAT 的配置。

```
NAT>enable
NAT #configure terminal
NAT(config)#access-list 1 permit 10.1.2.0 0.0.0.255
NAT(config)#ip nat pool natpool 192.168.1.3 192.168.1.10 prefix-length 24
NAT(config)#ip nat inside source list 1 pool natpool
NAT(config)#exit
NAT#write
```

(4) PAT 的配置。

NAT>enable
NAT #configure terminal
NAT(config)#access-list 2 permit 10.1.2.0 0.0.0.255
NAT(config)#ip nat inside source list 2 interface FastEthernet 1/0 overload
NAT(config)#exit
NAT#write

(5) 负载均衡的配置。

1) 在 NAT 交换机上增加的配置。

NAT>enable
NAT #configure terminal
NAT(config)#access-list 4 permit 192.168.2.50
NAT(config)#ip nat pool webserver 10.1.2.2 10.1.2.3 netmask 255.255.255.0 type rotary
NAT(config)#ip nat inside destination list 4 pool webserver
NAT(config)#exit
NAT#write

2) 在 Internet 交换机上增加的配置。

internet>enable
internet# configure terminal
internet (config)#ip route 192.168.2.0 255.255.255.0 192.168.1.1
internet (config)#exit
internet#write

## 七、实验测试和查看相关配置

(1) 静态 NAT 配置的测试。

在 Internet 交换机上 ping 192.168.1.100 的 IP 地址，显示结果如下：
internet#ping 192.168.1.100
Type escape sequence to abort.
Sending 5, 100-byte ICMP Echos to 192.168.1.100, timeout is 2 seconds:
!!!!!
Success rate is 100 percent (5/5), round-trip min/avg/max = 72/350/1224 ms

查看静态 NAT 转换记录。
Pro：协议的类型。
Inside global：内部全局地址。
Inside local：内部本地地址。
Outside local：外部本地地址。
Outside global：外部全局地址。
NAT#show ip nat translations

| Pro | Inside global | Inside local | Outside local | Outside global |
| --- | --- | --- | --- | --- |
| tcp | 192.168.1.100:80 | 10.1.2.2:80 | --- | --- #包转换条目 |

(2) 动态 NAT 配置的测试。

在 pc1 主机上 ping 192.168.1.2 的 IP 地址，显示结果如下：
pc1#ping 192.168.1.2
Type escape sequence to abort.
Sending 5, 100-byte ICMP Echos to 192.168.1.2, timeout is 2 seconds:

..!!!
Success rate is 60 percent (3/5), round-trip min/avg/max = 288/485/772 ms

查看动态 NAT 转换记录。

在 192.168.3.1 的交换机 ping 交换机 internet 交换机之后，动态 NAT 的转换记录如下：

NAT#show ip nat translations

| Pro | Inside global | Inside local | Outside local | Outside global | |
|---|---|---|---|---|---|
| icmp | 192.168.1.4:0 | 10.1.3.1:0 | 192.168.1.2:0 | 192.168.1.2:0 | |
| --- | 192.168.1.4 | 10.1.3.1 | --- | --- | |
| icmp | 192.168.1.3:7198 | 10.1.3.2:7198 | 192.168.1.2:7198 | 192.168.1.2:7198 | |
| icmp | 192.168.1.3:7199 | 10.1.3.2:7199 | 192.168.1.2:7199 | 192.168.1.2:7199 | |
| icmp | 192.168.1.3:7200 | 10.1.3.2:7200 | 192.168.1.2:7200 | 192.168.1.2:7200 | |
| icmp | 192.168.1.3:7201 | 10.1.3.2:7201 | 192.168.1.2:7201 | 192.168.1.2:7201 | |
| --- | 192.168.1.3 | 10.1.3.2 | --- | --- | #包转换条目 |

（3）PAT 配置的测试。

在 pc1 主机上 ping 192.168.1.2 的 IP 地址，显示结果如下：

pc1#ping 192.168.1.2
Type escape sequence to abort.
Sending 5, 100-byte ICMP Echos to 192.168.1.2, timeout is 2 seconds:
!!!!!
Success rate is 100 percent (5/5), round-trip min/avg/max = 140/451/1176 ms

查看 PAT 转换记录。

NAT(config)#do show ip nat translations

| Pro | Inside global | Inside local | Outside local | Outside global |
|---|---|---|---|---|
| icmp | 192.168.1.1:1 | 10.1.3.1:1 | 192.168.1.2:1 | 192.168.1.2:1 |
| icmp | 192.168.1.1:5718 | 10.1.3.2:5718 | 192.168.1.2:5718 | 192.168.1.2:5718 |
| icmp | 192.168.1.1:5719 | 10.1.3.2:5719 | 192.168.1.2:5719 | 192.168.1.2:5719 |
| icmp | 192.168.1.1:5720 | 10.1.3.2:5720 | 192.168.1.2:5720 | 192.168.1.2:5720 |
| icmp | 192.168.1.1:5721 | 10.1.3.2:5721 | 192.168.1.2:5721 | 192.168.1.2:5721 |
| icmp | 192.168.1.1:5722 | 10.1.3.2:5722 | 192.168.1.2:5722 | 192.168.1.2:5722 |

（4）负载均衡配置的测试。

通过浏览器访问 192.168.2.1 服务进行测试，第一次访问到到 web 服务器上，第二次访问到 web1 服务器上。结果如图 9-2 和图 9-3 所示。

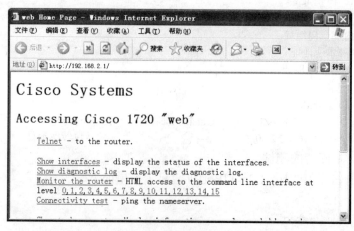

图 9-2　负载均衡测试 1

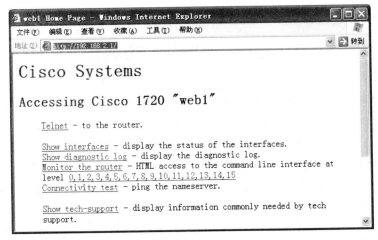

图 9-3　负载均衡测试 2

查看负载均衡的转换记录

NAT#　sho ip nat tr

| Pro Inside global | Inside local | Outside local | Outside global | |
|---|---|---|---|---|
| tcp 192.168.2.1:80 | 10.1.2.2:80 | 192.168.3.5:3285 | 192.168.3.5:3285 | #第一次访问转换到 10.1.2.2 的 web 上 |
| tcp 192.168.2.1:80 | 10.1.2.3:80 | 192.168.3.5:3286 | 192.168.3.5:3286 | #第二次访问转换到 10.1.2.3 的 web1 上 |

## 八、注意问题

在配置服务器负载均衡时，服务器的虚拟地址和 NAT 的 outside 端口不应在同一子网中。

# 实验十 DHCP 的配置

两台连接到互联网上的电脑相互之间通信，必须有各自的 IP 地址，但由于现在的 IP 地址资源有限，宽带接入运营商不能做到给每个报装宽带的用户都能分配一个固定的 IP 地址，所以要采用 DHCP 方式对上网的用户进行临时的地址分配。也就是你的电脑联上网，DHCP 服务器才从地址池里临时分配一个 IP 地址给你，每次上网分配的 IP 地址可能会不一样，这跟当时 IP 地址资源有关。当你下线的时候，DHCP 服务器可能就会把这个地址分配给之后上线的其他电脑。这样就可以有效节约 IP 地址，既保证了你的通信，又提高 IP 地址的使用率。

## 一、实验目的

理解 DHCP 服务器和 DHCP 中继的原理，掌握 DHCP 服务器和 DHCP 中继的配置。

## 二、实验内容

DHCP 服务器和 DHCP 中继的配置。

## 三、实验原理

DHCP（Dynamic Host Configure Protocol）是被广泛应用于 TCP/IP 协议的网络中的动态主机配置协议，使用 DHCP 协议有很多简单、易用的优点，主要表现在：网络管理员可以验证 IP 地址和其他配置参数，而不用去检查每个主机；DHCP 不会同时租借相同的 IP 地址给两台主机；DHCP 管理员可以约束特定的计算机使用特定的 IP 地址；可以为每个 DHCP 作用域设置很多选项；客户机在不同子网间移动时不需要重新设置 IP 地址。

（1）DHCP 协议的工作流程。

1) DHCP 客户机寻找 DHCP 服务器的阶段，可以称为发现阶段：DHCP 客户机以广播方式发送 DHCP discover 发现信息来寻找 DHCP 服务器，即向地址 255.255.255.255 发送特定的广播信息。网络上每一台安装了 TCP/IP 协议的主机都会接收到这种广播信息，但只有 DHCP 服务器才会做出响应。

2) DHCP 服务器提供 IP 地址的阶段，可以称为提供阶段：在网络中接收到 DHCP discover 发现信息的 DHCP 服务器都会做出响应，它从尚未出租的 IP 地址中挑选一个分配给 DHCP 客户机，向 DHCP 客户机发送一个包含出租的 IP 地址和其他设置的 DHCP offer 提供信息。

3) DHCP 客户机选择某台 DHCP 服务器提供的 IP 地址的阶段，可以称为选择阶段：如果有多台 DHCP 服务器向 DHCP 客户机发来的 DHCP offer 提供信息，则 DHCP 客户机只接受第一个收到的 DHCP offer 提供的信息，然后它就以广播方式回答一个 DHCP request 请求信息，该信息中包含向它所选定的 DHCP 服务器请求 IP 地址的内容。之所以要以广播方式回答，是为了通知所有的 DHCP 服务器，其将选择某台 DHCP 服务器所提供的 IP 地址。

4) DHCP 服务器确认所提供的 IP 地址的阶段，可以称为确认阶段：当 DHCP 服务器收到 DHCP 客户机回答的 DHCP request 请求信息之后，它便向 DHCP 客户机发送一个包含它所提供的 IP 地址和其他设置的 DHCP ack 确认信息，告诉 DHCP 客户机可以使用它所提供的 IP 地址。然后 DHCP 客户机便将其 TCP/IP 协议与网卡绑定，另外，除 DHCP 客户机选中的服务器外，其他的 DHCP 服务器都将收回曾提供的 IP 地址。

5）以后 DHCP 客户机每次重新登录网络时，就不需要再发送 DHCP discover 发现信息了，而是直接发送包含前一次所分配的 IP 地址的 DHCP request 请求信息。当 DHCP 服务器收到这一信息后，它会尝试让 DHCP 客户机继续使用原来的 IP 地址，并回答一个 DHCP ack 确认信息。如果此 IP 地址已无法再分配给原来的 DHCP 客户机使用时（比如此 IP 地址已分配给其他 DHCP 客户机使用），则 DHCP 服务器给 DHCP 客户机回答一个 DHCP nack 否认信息。当原来的 DHCP 客户机收到此 DHCP nack 否认信息后，它就必须重新发送 DHCP discover 发现信息来请求新的 IP 地址。

6）DHCP 服务器向 DHCP 客户机出租的 IP 地址一般都有一个租借期限，期满后 DHCP 服务器便会收回出租的 IP 地址。如果 DHCP 客户机要延长其 IP 租约，则必须更新其 IP 租约。DHCP 客户机启动时和 IP 租约期限过一半时，DHCP 客户机都会自动向 DHCP 服务器发送更新其 IP 租约的信息。如果该客户机在租约规定的时间内一直占用 DHCP 分配的地址并且在线，则该地址将一直被该客户机所使用。

（2）DHCP 中继。

DHCP 客户使用 IP 广播来寻找同一网段上的 DHCP 服务器。当服务器和客户段处在不同网段（即被路由器分割开来）时，路由器是不会转发这样广播包的。因此可能需要在每个网段上设置一个 DHCP 服务器，虽然 DHCP 只消耗很小的一部分资源，但多个 DHCP 服务器毕竟会带来管理上的不方便。DHCP 中继的使用使一个 DHCP 服务器同时为多个网段服务成为可能。

为了让路由器可以帮助转发广播请求数据包，使用 ip help-address 命令。通过使用该命令，路由器可以配置为接受广播请求，然后将其以单播方式转发指定 IP 地址。默认情况下，ip help-address 转发 8 种 UDP 服务：Time、Tacacs、DNS、BOOTP/DHCP 服务器、BOOTP/DHCP 客户、TFTP、NetBios 名称服务和 NetBios 数据报服务。

在 DHCP 广播情况下，客户在本地网段广播一个 DHCP 发现分组。网关获得这个分组，如果配置了帮助地址，网关就将 DHCP 分组通过单播方式转发到特定地址。

交换机或路由器上配置 ip help-address 命令后，默认情况下，路由器不仅转发 DHCP 请求，同时也转发其他的 UDP 报，这样很可能会增加 DHCP 服务器所在链路的负担，同时也增加了 DHCP 服务器的 CPU 利用率，这可能会引起很严重的网络通信问题。所以要通过相关命令禁止相关信息的转发。

### 四、实验需要掌握的命令

Switch(config)#service dhcp
#开启 DHCP 服务，此命令默认为开启的
Switch(config)#ip dhcp pool *poolname*
#定义名称为 *poolname* 的 DHCP 地址池
Switch(dhcp-config)#network *Network-number* {*mask* | */nn*}
#用 network 命令来定义网络地址的范围，*Network-number* 为网段地址，*mask* 和*/nn* 二选一，*mask* 为子网掩码，*/nn* 为子网掩码的长度
Switch (dhcp-config)#default-router *IP-ADDRESS*
#定义要分配的网关地址，*IP-ADDRESS* 为网关 IP 地址
Switch (dhcp-config)#dns-server *IP-ADDRESS*

#定义要分配的域名解析服务器地址，IP-ADDRESS 为域名解析服务器 IP 地址
Switch(dhcp-config)#host *IP-ADDRESS* {*mask* | */nn*}
#保留主机的地址，不用于动态分配
Switch(config-if)#ip helper-address *IP-ADDRESS*
#指定 DHCP 服务器的地址，表示通过当前端口向该 DHCP 服务器发送 DHCP 请求包，*IP-ADDRESS* 为 DHCP 服务器 IP 地址
Switch(config-if)#ip address dhcp client-id Ethernet0
#开启 DHCP 的客户端，以使得该接口动态地从 DHCP 服务器端获得 IP 地址
Switch(config)#ip dhcp excluded-address *start-IP-ADDRESS end-IP-ADDRESS*
#该范围内的 IP 地址不能分配给客户端，*start-IP-ADDRESS* 为起始 IP 地址，*end-IP-ADDRESS* 为结束 IP 地址
Switch(config)#no ip forward-protocol udp tftp
#禁止转发 tftp 请求数据报文
Switch(config)#no ip forward-protocol udp nameserver
#禁止转发 nameserver 请求数据报文
Switch(config)#no ip forward-protocol udp domain
#禁止转发 domain 请求数据报文
Switch(config)#no ip forward-protocol udp time
#禁止转发 time 请求数据报文
Switch(config)#no ip forward-protocol udp netbios-ns
#禁止转发 netbios-ns 请求数据报文
Switch(config)#no ip forward-protocol udp netbios-dgm
#禁止转发 netbios-dgm 请求数据报文
Switch(config)#no ip forward-protocol udp tacacs
#禁止转发 tacacs 请求数据报文
Switch#show ip interface
#显示相关的 ip help-address 配置信息

## 五、实验拓扑图和拓扑文件

（1）网络拓扑图。
网络拓扑图如图 10-1 所示。
（2）网络文件。
autostart = false
ghostios = true
sparsemem = true

[localhost]

  [[3660]]
   image = C:\Program Files\Dynamips\images\c3660.bin
   ram = 160

实验十 DHCP 的配置

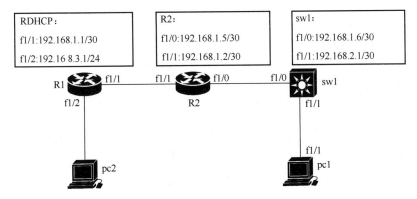

图 10-1 网络拓扑图

```
[[ROUTER RDHCP]]
    idlepc = 0x6064179c
    slot1 = NM-16ESW
    model = 3660
    f1/1 = R2 f1/1
    f1/2 = pc2 f1/2

[[router R2]]
    idlepc = 0x6064179c
    model = 3660
    slot1 = NM-16ESW
    f1/0 = sw1 f1/0

[[router sw1]]
    idlepc = 0x6064179c
    slot1 = NM-16ESW
    model = 3660
    f1/1 = pc1 f1/1

[[router pc1]]
    idlepc = 0x6064179c
    slot1 = NM-16ESW
    model = 3660

[[router pc2]]
    idlepc = 0x6064179c
    slot1 = NM-16ESW
    model = 3660
```

六、实验配置步骤

（1）路由器 RDHCP 的配置。

```
router>enable
router#configure terminal
router(config)#hostname DHCP
DHCP(config)#service dhcp
DHCP(config)#ip dhcp pool office1
DHCP(dhcp-config)#network 192.168.2.0 255.255.255.0
```

```
DHCP(dhcp-config)#dns-server 192.168.1.254
DHCP(dhcp-config)#default-router 192.168.2.1
DHCP(dhcp-config)#exit
DHCP(config)#ip dhcp pool office
DHCP(dhcp-config)#network 192.168.3.0 255.255.255.0
DHCP(dhcp-config)#dns-server 192.168.1.254
DHCP(dhcp-config)#default-router 192.168.3.1
DHCP(dhcp-config)#exit
DHCP(config)#interface FastEthernet1/1
DHCP(config-if)#no switchport
DHCP(config-if)#ip address 192.168.1.1 255.255.255.252
DHCP(config-if)#duplex full
DHCP(config-if)#speed 100
DHCP(config-if)#interface FastEthernet1/2
DHCP(config-if)#no switchport
DHCP(config-if)#ip address 192.168.3.1 255.255.255.252
DHCP(config-if)#duplex full
DHCP(config-if)#speed 100
DHCP(config-if)#exit
DHCP(config)#ip route 192.168.1.4 255.255.255.252 192.168.1.2
DHCP(config)#ip route 192.168.2.0 255.255.255.0 FastEthernet1/1
DHCP(config)#do write
```

（2）路由器 R2 的配置。

```
router>enable
router#configure terminal
router(config)#hostname R2
R2(config)#interface FastEthernet1/0
R2(config-if)#no switchport
R2(config-if)#ip dhcp relay information trusted
R2(config-if)#ip address 192.168.1.5 255.255.255.252
R2(config-if)#duplex full
R2(config-if)#speed 100
R2(config-if)#interface FastEthernet1/1
R2(config-if)#no switchport
R2(config-if)#ip address 192.168.1.2 255.255.255.252
R2(config-if)#duplex full
R2(config-if)#speed 100
R2(config-if)#ip route 192.168.2.0 255.255.255.0 192.168.1.6
R2(config-if)#ip route 192.168.3.0 255.255.255.0 192.168.1.1
R2(config-if)#do write
```

（3）交换机 sw1 的配置。

```
switch>enable
switch#configure terminal
switch(config)#hostname sw1
sw1(config)#service dhcp
sw1(config)#interface FastEthernet1/0
sw1(config-if)#no switchport
sw1(config-if)#ip dhcp relay information trusted
sw1(config-if)#ip address 192.168.1.6 255.255.255.252
sw1(config-if)#duplex full
```

sw1(config-if)#speed 100
sw1(config-if)#interface FastEthernet1/1
sw1(config-if)#no switchport
sw1(config-if)#ip address 192.168.2.1 255.255.255.0
sw1(config-if)#ip helper-address 192.168.1.1
sw1(config-if)#duplex full
sw1(config-if)#speed 100
sw1(config-if)#exit
sw1(config)#ip route 192.168.1.0 255.255.255.252 192.168.1.5
sw1(config)#ip route 192.168.3.0 255.255.255.0 192.168.1.5
s sw1(config)#do write

（4）主机 pc1 的配置。

switch>enable
switch #configure terminal
switch (config)#hostname pc1
pc1(config)#interface FastEthernet1/1
pc1(config-if)#no switchport
pc1(config-if)#ip address dhcp client-id FastEthernet1/1
pc1(config-if)#duplex full
pc1(config-if)#speed 100
pc1(config-if)#end
pc1#write

（5）主机 pc2 的配置。

switch>enable
switch#configure terminal
switch(config)#hostname pc2
pc2(config)#interface FastEthernet1/2
pc2(config-if)#no switchport
pc2(config-if)#ip address dhcp client-id FastEthernet1/2
pc2(config-if)#end
pc2#write

## 七、实验测试和查看相关配置

（1）查看 pc1 的端口状态。

pc1#show interfaces f1/1
FastEthernet1/1 is up, line protocol is up
　　Hardware is FastEthernet, address is cc03.0e78.f101 (bia cc03.0e78.f101)
　　Internet address is **192.168.2.2/24**
　　MTU 1500 bytes, BW 100000 Kbit, DLY 100 usec,
　　　　reliability 255/255, txload 1/255, rxload 1/255
　　Encapsulation ARPA, loopback not set
　　Keepalive set (10 sec)
　　Full-duplex, 100Mb/s
……

（2）查看 pc2 的端口信息。

Router#sho int f1/2
FastEthernet1/2 is up, line protocol is up
　　Hardware is FastEthernet, address is cc04.0e78.f102 (bia cc04.0e78.f102)
　　Internet address is 192.168.3.2/24

MTU 1500 bytes, BW 100000 Kbit, DLY 100 usec,
　　reliability 255/255, txload 1/255, rxload 1/255
Encapsulation ARPA, loopback not set
Keepalive set (10 sec)
Auto-duplex, Auto-speed
ARP type: ARPA, ARP Timeout 04:00:00
……

（3）查看交换机 sw1 的端口状态。
Router#sho ip int f1/1
FastEthernet1/1 is up, line protocol is up
　Internet address is 192.168.2.1/24
　Broadcast address is 255.255.255.255
　Address determined by non-volatile memory
　MTU is 1500 bytes
　**Helper address is 192.168.1.1**
……

（4）查看路由器 DHCP 的端口状态。
DHCP#sho ip int f1/2
FastEthernet1/2 is up, line protocol is up
　Internet address is 192.168.3.1/24
　Broadcast address is 255.255.255.255
　Address determined by setup command
　MTU is 1500 bytes
　**Helper address is not set**

## 八、注意问题

在配置模拟 PC 路由器的端口自动获得 IP 地址时，如果在接口模式下没有 ip address dhcp client-id FastEthernet1/1 命令，则可以先将端口设置为二层接口，然后再设置为三层接口，最后再输入 ip address dhcp client-id FastEthernet1/1 命令。

# 实验十一　VRRP 的配置

VRRP 是一种容错协议，它通过把几台路由设备联合组成一台虚拟的路由设备，并通过一定的机制来保证当主机的下一跳设备出现故障时，可以及时将业务切换到备份设备，从而保持通信的连续性和可靠性。

## 一、实验目的

（1）理解 VRRP 的原理；
（2）掌握 VRRP 的配置；
（3）掌握 VRRP 和 VTP 综合应用配置。

## 二、实验内容

（1）掌握 VRRP 的配置；
（2）掌握 VRRP 和 VTP 综合应用配置。

## 三、VRRP 协议简介

（1）协议概述。

在基于 TCP/IP 协议的网络中，为了保证不直接物理连接的设备之间的通信，必须指定路由。目前常用的指定路由的方法有两种：一种是通过路由协议（比如内部路由协议 RIP 和 OSPF）动态学习；另一种是静态配置。在每一个终端都运行动态路由协议是不现实的，大多客户端操作系统平台都不支持动态路由协议，即使支持也受到管理开销、收敛度、安全性等许多问题的限制。因此普遍采用对终端 IP 设备静态路由配置，一般是给终端设备指定一个或者多个默认网关（Default Gateway）。静态路由的方法简化了网络管理的复杂度，减轻了终端设备的通信开销，但是它仍然有一个缺点：如果作为默认网关的路由器损坏，所有使用该网关为下一跳主机的通信必然要中断。即便配置了多个默认网关，如不重新启动终端设备，也不能切换到新的网关。采用虚拟路由冗余协议（Virtual Router Redundancy Protocol，VRRP）可以很好地避免静态指定网关的缺陷。

在 VRRP 协议中，有两组重要的概念：VRRP 路由器和虚拟路由器，主控路由器和备份路由器。VRRP 路由器是指运行 VRRP 的路由器，是物理实体；虚拟路由器是指 VRRP 协议创建的，是逻辑概念。一组 VRRP 路由器协同工作，共同构成一台虚拟路由器。该虚拟路由器对外表现为一个具有唯一固定 IP 地址和 MAC 地址的逻辑路由器。处于同一个 VRRP 组中的路由器具有两种互斥的角色：主控路由器和备份路由器，一个 VRRP 组中有且只有一台处于主控角色的路由器，可以有一个或者多个处于备份角色的路由器。VRRP 协议使用选择策略从路由器组中选出一台作为主控，负责转发 ARP 数据包和 IP 数据包，组中的其他路由器作为备份的角色处于待命状态。当由于某种原因主控路由器发生故障时，备份路由器能在几秒钟的时延后升级为主路由器。由于此切换非常迅速而且不用改变 IP 地址和 MAC 地址，故对终端使用者系统是透明的。

（2）工作原理。

一个 VRRP 路由器有唯一的标识——VRID，范围为 0～255。该路由器对外表现为唯一的虚拟 MAC 地址，地址的格式为 00-00-5E-00-01-[VRID]。主控路由器负责对 ARP 请求用该 MAC 地址做应答。这样，无论如何切换，保证给终端设备的是唯一一致的 IP 和 MAC 地址，减少了切换对终端设备的影响。

VRRP 控制报文只有一种：VRRP 通告（advertisement）。它使用 IP 多播数据包进行封装，组地址为 224.0.0.18，发布范围只限于同一局域网内。这保证了 VRID 在不同网络中可以重复使用。为了减少网络带宽消耗，只有主控路由器才可以周期性地发送 VRRP 通告报文。备份路由器在连续三个通告间隔内收不到 VRRP，或收到优先级为 0 的通告后启动新一轮 VRRP 选举。

在 VRRP 路由器组中，按优先级选举主控路由器，VRRP 协议中优先级范围是 0～255。若 VRRP 路由器的 IP 地址和虚拟路由器的接口 IP 地址相同，则称该虚拟路由器为 VRRP 组中的 IP 地址所有者；IP 地址所有者自动具有最高优先级 255。优先级 0 一般用在 IP 地址所有者主动放弃主控者角色时使用。可配置的优先级范围为 1～254。优先级的配置原则可以依据链路的速度和成本、路由器性能和可靠性以及其他管理策略设定。主控路由器的选举中，高优先级的虚拟路由器获胜，因此，如果在 VRRP 组中有 IP 地址所有者，则它总是作为主控路由的角色出现。对于相同优先级的候选路由器，按照 IP 地址大小顺序选举。VRRP 还提供了优先级抢占策略，如果配置了该策略，高优先级的备份路由器便会剥夺当前低优先级的主控路由器而成为新的主控路由器。

为了保证 VRRP 协议的安全性，提供了两种安全认证措施：明文认证和 IP 头认证。明文认证方式要求：在加入一个 VRRP 路由器组时，必须同时提供相同的 VRID 和明文密码。适合于避免在局域网内的配置错误，但不能防止通过网络监听方式获得密码。IP 头认证的方式提供了更高的安全性，能够防止报文重放和修改等攻击。

### 四、实验需要掌握的命令

Router(config-if)# vrrp *group-number* ip *IP-address*
#配置 VRRP 组和虚拟路由器的 IP 地址，其中 *group-number* 为组号，*IP-address* 为 IP 地址

Router(config-if)#vrrp *group-number* priority *priority-level*
#配置优先级，其中 *priority-level* 为优先级，范围是 1～254

Router(config-if)#vrrp *group-number* preempt
#配置为抢占模式

Router(config-if)#vrrp *group-number* timer advertise [msec] *number*
#配置 VRRP 通告时间，默认为 1 秒，msec 是将单位设置为毫秒

Router(config-if)#vrrp *group-number* authentication md5 key-string [0|7] *password* timeout *number*
#设置 MD5 认证，*password* 为密码，*number* 为超时的时间，0 表示不以加密方式显示密码，7 表示以加密的方式显示密码

Router(config-if)#vrrp *group-number* authentication md5 key-chain key *chain-name*
#设置 MD5 钥匙串认证，*chain-name* 为密钥链的名称，必须事先定义密钥链

Router(config-if)#vrrp *group-number* authentication text *password*

#设置明文认证，password 为密码
Router(config)#track *track-object* interface *interface-typ number* {line-protocol | ip routing}
#设置端口追踪
Router(config-if)#vrrp *group-number* track *track-object* decrement *priority-level*
#在端口上应用追踪，*priority-level* 为追踪的端口出现故障时，减少 VRRP 组的优先级的值

Router#show vrrp brief
#显示 VRRP 的信息

Router#show vrrp interface *interface-typ number*
#显示一个端口上的 VRRP 信息

Router#show track
#显示追踪的信息

## 五、实验拓扑图和网络文件

（1）网络拓扑图。

网络拓扑图如图 11-1 所示。

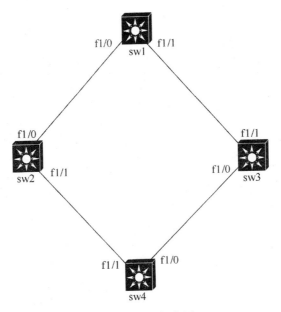

图 11-1　网络拓扑图

表 11-1　IP 地址规划

| 设备名称 | 端口号 | IP 地址 | 备注 |
| --- | --- | --- | --- |
| sw1 | vlan4 | 192.168.1.1/29 | 包含 f1/0、f1/1 端口 |
| sw2 | vlan2 | 192.168.2.2/24 | valn2 的 VRRP 组号为 1，虚拟地址为 192.168.2.1/29；valn3 的 VRRP 组号为 2，虚拟地址为 192.168.3.1/29；f1/1 配置为 Trunk 模式；f1/0 配置为 Access 模式，属于 vlan4 |
|  | vlan3 | 192.168.3.2/24 |  |
|  | vlan4 | 192.168.1.2/29 |  |

续表

| 设备名称 | 端口号 | IP 地址 | 备注 |
|---|---|---|---|
| sw3 | vlan2 | 192.168.2.3/24 | valn2 的 VRRP 组号为 1，虚拟地址为 192.168.2.1/29；valn3 的 VRRP 组号为 2，虚拟地址为 192.168.3.1/29；f1/0 配置为 Trunk 模式；f1/1 配置为 Access 模式，属于 vlan4 |
| | vlan3 | 192.168.3.3/24 | |
| | vlan4 | 192.168.1.3/29 | |
| sw4 | vlan2 | | vlan2 包含 f1/2-8；vlan2 包含 f1/9-15；f1/0 与 f1/1 配置为 Trunk 模式 |
| | vlan3 | | |

（2）网络文件。

```
#VRRP.NET 文件开始
# VRRP lab
autostart = false
ghostios = true
sparsemem = true

[localhost]

    [[3660]]
    image = ..\c3660-is.bin
    ram = 160

    [[ROUTER sw1]]
    idlepc = 0x60719b98
    model = 3660
    slot1 = NM-16ESW
    f1/0 = sw2 f1/0
    f1/1 = sw3 f1/1

    [[router sw2]]
    idlepc = 0x60719b98
    model = 3660
    slot1 = NM-16ESW
    f1/1 = sw4 f1/1

    [[router sw3]]
    idlepc = 0x60719b98
    model = 3660
    slot1 = NM-16ESW
    f1/0 = sw4 f1/0

    [[router sw4]]
    idlepc = 0x60719b98
    model = 3660
    slot1 = NM-16ESW
#VRRP.NET 文件结束
```

## 六、实验配置步骤

（1）交换机 sw1 的配置步骤。

switch>enable
switch#vlan database
switch(vlan)#vlan 4      #创建 vlan4
switch(vlan)#exit
switch#configure terminal
switch(config)#hostname sw1
sw1(config-if-range)#interface range FastEthernet1/0 - 1
sw1(config-if-range)#switchport
sw1(config-if-range)#switchport access vlan 4
sw1(config-if-range)#speed 100
sw1(config-if-range)#duplex full
sw1(config-if-range)#no shutdown
sw1(config-if-range)#interface Vlan4
sw1(config-if)#ip address 192.168.1.1 255.255.255.248
sw1(config-if)#no shutdown
sw1(config-if)#exit
sw1(config)#ip route 192.168.2.0 255.255.254.0 Vlan4    #配置到网络 192.168.2.0/24 和 192.168.3.0/24 的路由
sw1(config)#do write

（2）交换机 sw2 的配置步骤。

switch>enable
switch#vlan database
switch(vlan)#vlan 2
switch(vlan)#vlan 3
switch(vlan)#vlan 4
#上面的命令是创建 vlan
switch(vlan)#exit
switch#configure terminal
switch(config)#hostname sw2
sw2(config)#track 1 interface Vlan4 line-protocol   #设置端口追踪
sw2(config)#interface FastEthernet1/0
sw2(config-if)#switchport
sw2(config-if)#switchport access vlan 4
sw2(config-if)#speed 100
sw2(config-if)#duplex full
sw2(config-if)#no shutdown
sw2(config-if)#interface FastEthernet1/1
sw2(config-if)#switchport
sw2(config-if)#switchport mode trunk
sw2(config-if)#switchport trunk allowed vlan all
sw2(config-if)#switchport trunk encapsulation dot1q
sw2(config-if)#speed 100
sw2(config-if)#duplex full
sw2(config-if)#no shutdown

```
sw2(config-if)#interface Vlan2
sw2(config-if)#ip address 192.168.2.2 255.255.255.0
sw2(config-if)#vrrp 1 ip 192.168.2.1
sw2(config-if)#vrrp 1 ip 192.168.2.2 secondary
sw2(config-if)#vrrp 1 priority 100
sw2(config-if)#vrrp 1 track 1 decrement 50 #将上面的端口追踪应用到 VRRP 1 组
#上面四条命令基于 vlan2 设置 vrrp 的配置
sw2(config-if)#no shutdown
sw2(config-if)#interface Vlan3
sw2(config-if)#ip address 192.168.3.2 255.255.255.0
sw2(config-if)#vrrp 2 ip 192.168.3.1
sw2(config-if)#vrrp 2 ip 192.168.3.2 secondary
sw2(config-if)#vrrp 2 priority 60
sw2(config-if)#vrrp 2 authentication md5 key-string abcd
#上面四条命令基于 vlan3 设置 vrrp 的配置
sw2(config-if)#no shutdown
sw2(config-if)#interface Vlan4
sw2(config-if)#ip address 192.168.1.2 255.255.255.248
sw2(config-if)#no shutdown
sw2(config-if)#do write
```

（3）交换机 sw3 的配置步骤。

```
switch>enable
switch#vlan database
switch(vlan)#vlan 2
switch(vlan)#vlan 3
switch(vlan)#vlan 4
#上面的命令是创建 vlan
switch(vlan)#exit
switch#configure terminal
switch(config)#hostname sw3
sw3(config-if)#track 1 interface Vlan4 line-protocol
sw3(config-if)#interface FastEthernet1/0
sw3(config-if)#switchport
sw3(config-if)#switchport mode trunk
sw3(config-if)#switchport trunk allowed vlan all
sw3(config-if)#switchport trunk encapsulation dot1q
sw3(config-if)#speed 100
sw3(config-if)#duplex full
sw3(config-if)#no shutdown
sw3(config-if)#interface FastEthernet1/1
sw3(config-if)#switchport
sw3(config-if)#switchport access vlan 4
sw3(config-if)#speed 100
sw3(config-if)#duplex full
sw3(config-if)#no shutdown
sw3(config-if)#interface Vlan2
sw3(config-if)#ip address 192.168.2.3 255.255.255.0
```

sw3(config-if)#vrrp 1 ip 192.168.2.1
sw3(config-if)#vrrp 1 ip 192.168.2.3 secondary
sw3(config-if)#vrrp 1 priority 60
#上面四条命令基于 vlan2 设置 VRRP 的配置
sw3(config-if)#no shutdown
sw3(config-if)#interface Vlan3
sw3(config-if)#ip address 192.168.3.3 255.255.255.0
sw3(config-if)#vrrp 2 ip 192.168.3.1
sw3(config-if)#vrrp 2 ip 192.168.3.3 secondary
sw3(config-if)#vrrp 2 priority 100
sw3(config-if)#vrrp 2 authentication md5 key-string abcd
sw3(config-if)#vrrp 2 track 1 decrement 50    #将上面的端口追踪应用到 VRRP 2 组
#上面五条命令基于 vlan3 设置 VRRP 的配置
sw3(config-if)#no shutdown
sw3(config-if)#interface Vlan4
sw3(config-if)#ip address 192.168.1.3 255.255.255.248
sw3(config-if)#no shutdown
sw3(config-if)#do write

（4）交换机 sw4 的配置步骤。

switch>enable
switch#vlan database
switch(vlan)#vlan 2
switch(vlan)#vlan 3
#上面的命令是创建 vlan
switch(vlan)#exit
switch#configure terminal
switch(config)#hostname sw4
sw4(config)#interface range FastEthernet1/0 - 1
sw4(config-if-range)#switchport
sw4(config-if-range)#switchport mode trunk
sw4(config-if-range)#switchport trunk allowed vlan all
sw4(config-if-range)#switchport trunk encapsulation dot1q
sw4(config-if-range)#speed 100
sw4(config-if-range)#duplex full
sw4(config-if-range)#no shutdown
w4(config-if-range)#exit
w4(config)#no ip routing
w4(config)#write

## 七、实验测试和查看相关配置

（1）测试直连端口。

测试每一对直连端口通信是否正常，例如：在主机 sw1 上 ping 路由器 sw2 的 vlan4 端口，结果如下，说明这一对直连端口通信正常。如果没有收到任何 ICMP 的应答数据包，则说明这一对直连端口通信不正常，这时要查看端口的 IP 地址、子网掩码、speed、duplex 等参数的配置。

sw1>ping 192.168.1.2

Type escape sequence to abort.
Sending 5, 100-byte ICMP Echos to 192.168.1.2, timeout is 2 seconds:
!!!!!
Success rate is 100 percent (5/5), round-trip min/avg/max = 152/202/260 ms

（2）查看 sw2 和 sw3 的 VRRP 配置。

sw2#show vrrp all
Vlan2 - Group 1    #vlan2 隶属于 VRRP 组 1
  State is Master    #在 VRRP 组 1 中，当前设备中的 vlan2 为主路由
  Virtual IP address is 192.168.2.1    #vlan2 的虚拟网关的 IP 地址为 192.168.2.1
    Secondary Virtual IP address is 192.168.2.2    #vlan2 的虚拟网关的第二个 IP 地址为 192.168.2.2
  Virtual MAC address is 0000.5e00.0101    #vlan2 的虚拟 MAC 地址为 0000.5e00.0101
  Advertisement interval is 1.000 sec    #宣告时间间隔为 1 秒
  Preemption enabled    #VRRP 组 1 的模式为抢占模式
  Priority is 100    #VRRP 组 1 的优先级为 100
    Track object 1 state Up decrement 50    #追踪目标 1，当追踪的端口线路协议为 up；当追踪的端口线路协议为 down 时，当前设备 VRRP 组 1 的优先级减 50
  Master Router is 192.168.2.2 (local), priority is 100    #当前交换机在 VRRP 组 1 中为主路由，IP 地址为 192.168.2.2，优先级为 100
  Master Advertisement interval is 1.000 sec
  Master Down interval is 3.609 sec

Vlan3 - Group 2
  State is Backup
  Virtual IP address is 192.168.3.1
    Secondary Virtual IP address is 192.168.3.2
  Virtual MAC address is 0000.5e00.0102
  Advertisement interval is 1.000 sec
  Preemption enabled
  Priority is 60
  Authentication MD5, key-string "abcd"
  Master Router is 192.168.3.3, priority is 100
  Master Advertisement interval is 1.000 sec
  Master Down interval is 3.765 sec (expires in 3.501 sec)

sw3#show vrrp all
Vlan2 - Group 1
  State is Backup
  Virtual IP address is 192.168.2.1
    Secondary Virtual IP address is 192.168.2.3
  Virtual MAC address is 0000.5e00.0101
  Advertisement interval is 1.000 sec
  Preemption enabled
  Priority is 60
  Master Router is 192.168.2.2, priority is 100
  Master Advertisement interval is 1.000 sec

Master Down interval is 3.765 sec (expires in 3.145 sec)

Vlan3 - Group 2
  State is Master
  Virtual IP address is 192.168.3.1
    Secondary Virtual IP address is 192.168.3.3
  Virtual MAC address is 0000.5e00.0102
  Advertisement interval is 1.000 sec
  Preemption enabled
  Priority is 100
    Track object 1 state Up decrement 50
  Authentication MD5, key-string "abcd"
  Master Router is 192.168.3.3 (local), priority is 100
  Master Advertisement interval is 1.000 sec
  Master Down interval is 3.609 sec

（3）全网测试。

在 sw4 上接上一台主机，如果连接的端口属于 vlan2，则 IP 地址一定是 192.168.2.0/24 网段的地址，如果连接的端口属于 vlan3，IP 地址一定是 192.168.3.0/24 网段的地址；然后在 PC 上 ping 192.168.1.1，看是否连通。如果不通，可能是路由的配置问题，也可能是端口的配置问题；如果是路由的问题，通过 show ip route 查看路由配置。如果路由没有问题，用 show interface *type number* 查看端口的配置，端口的配置要区分 Trunk 模式和 Access 模式，不能配置错误，否则也会导致网络不通。

## 八、注意问题

（1）端口的模式配置不能配置错误，是 Trunk 模式就不能配置为 Access 模式。

（2）端口的追踪只在 master 网关上设置，在 backup 网关上不用设置。

（3）本实验在配置冗余网关的同时也配置了网络负载均衡。

（4）虚拟 MAC 地址不能和相应的端口绑定，即在配置文件中不能出现与虚拟端口相关 mac-address-table static 0000.0c07.ac01 interface FastEthernet1/1 vlan 2 命令行，否则就无法 ping 通此网关。如果出现此命令行，可以用 no mac-address-table static 0000.0c07.ac01 将此命令行删除。其中 0000.0c07.ac01 为 VRRP1 组默认虚拟 MAC 地址。

# 实验十二  HSRP 的配置

HSRP 是一种容错协议，它通过把几台路由设备联合组成一台虚拟的路由设备，并通过一定的机制来保证当主机的下一跳设备出现故障时，可以及时将业务切换到备份设备，从而保持通信的连续性和可靠性。

## 一、实验目的

（1）理解 HSRP 的原理和作用；
（2）掌握 HSRP 的配置。

## 二、实验内容

在企业网络中利用 HSRP 配置冗余网关，提高网络的可靠性，同时实现网络流量的负载均衡。

## 三、实验原理

（1）介绍。

热备份路由协议（HSRP）主要是向我们提供了一种机制，它的设计目的主要在于支持 IP 传输失败情况下的不间断服务。具体地说，就是本协议用于在源主机无法动态地学习到首跳路由器 IP 地址的情况下防止首跳路由的失败。它主要用于多接入、多播和广播局域网（例如以太网）。当然 HSRP 并不是有意要取代现有的动态路由发现机制，而这些现有的路由协议仍可以继续使用，只不过不是在任何可能的情况下。以前的大部分主机都不支持动态路由发现协议，它们是通过配置默认路由来进行工作的。而 HSRP 却为它们提供了一种失败服务机制，在 HSRP 中所涉及到的所有路由器都被假设为已经配好了合适的 IP 路由协议，并且也已经存在了若干条路由。

在使用 HSRP 时，一组路由器的工作将一致地表现为局域网上通往主机的一个虚拟路由器的工作，这组路由器就称为一个 HSRP 组或备份组。这个组中将选出一个路由器来负责转发由主机发给虚拟路由器的数据包，这个路由器就是所谓的活动路由器。另一台路由器将被选为备份路由器。在活动路由器失效的情况下，备份路由器将承担活动路由器的包转发功能。即使可以任意制定运行 HSRP 的路由器的数量，但只有活动路由器才能转发发送给虚拟路由器的数据包。

为了把网络阻塞降到最低限度，网络中只有活动路由器和备份路由器可以在完成 HSRP 协议选择过程后，发送一次 HSRP 消息包。如果活动路由器失效，则备份路由器将取代它作为新的活动路由器工作。而当备份路由器失效或者它变成了活动路由器时，另外一个路由器将被选为备份路由器。

在某个局域网里，多个热备份组可以共存和重叠。每个备份组都仿效一个虚拟路由器。对于每个备份组来说，都有一个为别人所知的 MAC 地址和一个 IP 地址。而这个 IP 地址应该是这个局域网中第一个子网中的地址，但必须不同于设置在所有路由器端口上的地址和局域网中主机的地址，甚至包括为其他 HSRP 组设的地址。

如果在一个局域网中设置了多个 HSRP 组，那么分配主机通过指定其中一个备份组为网关，就可以对网络流量产生负载均衡。

（2）相关概念。

活动路由器：当前代表虚拟路由器转发数据包的路由器。

备份路由器：第一备份路由器。

备份组：参与到 HSRP 中，用以仿效虚拟路由器的一组路由器。

Hellotime：一个给定路由器成功地发出两个 HSRP hello 消息包之间的间隔。

Holdtime：假定发送路由器失败的情况下，收到两个 hello 消息包之间的间隔。

（3）协议介绍。

在备份组里，路由器通过发送各种不同的消息，周期性地广播状态信息。

备份协议运行在 UDP 层上，使用 1985 端口号。包发送多播地址 224.0.0.2，TTL 为 1。在 HSRP 包格式里，路由器使用它们的真实 IP 地址作为源地址，而不使用虚拟地址。这对于使 HSRP 路由器能够准确定义彼此是非常重要的。

下面是 UDP 帧格式的数据部分的格式：

| 0 1 2 3 4 5 6 7 | 8 9 0 1 2 3 4 5 | 6 7 8 9 0 1 2 3 | 4 5 6 7 8 9 0 1 |
|---|---|---|---|
| 版本号 | 操作码 | 状态 | Hellotime |
| Holdtime | 优先级 | 组 | 保留 |
| 授权数据 | | | |
| 授权数据 | | | |
| 虚拟IP地址 | | | |

版本号：1 个字节，HSRP 信息的版本号，本文所描述的版本号为 0。

操作码：1 个字节，操作码说明的是包含在这个包里的信息类型，可能的值有：0 - Hello、1 - Coup、2 - Resign。

Hello 类型消息是用来表明路由器正在工作，并且有能力成为活动路由器或者备份路由器。

Coup 类型消息是在当一个路由器希望变成活动路由器时才被发送的信息。

Resign 类型消息则是当一个路由器不希望再做活动路由器时才被发送的信息。

状态：1 个字节，在备份组中的每个路由器都在运行着一个状态机制。这个状态域描述的是发送消息的路由器的当前状态。可能的状态值有：0 - Initial、1 - Learn、2 - Listen、4 - Speak、8 - Standby、16 - Active。

Hellotime：1 个字节，这个域在 Hello 消息中是非常有意义的。它包含了路由器发送 Hello 消息的大约的间隔时间，这个时间是以秒为单位。如果路由器上没有配置 Hellotime，那么它将会向活动路由器发送的 Hello 消息学习。

而如果 Hellotime 没有被设置，而且 Hello 消息已经被授权，则只能通过学习来获取 Hellotime。发送 Hello 消息的路由器必须引入 Hello 消息中的 Hellotime 域使用的 Hellotime 值。

如果没有从活动路由器发过来的 Hello 消息中学习到 Hellotime，并且也没有手工配置 Hellotime，那么将把它的值默认地定为 3 秒钟。

Holdtime：1 个字节，这个域只在 Hello 消息中有效。它标明了当前的 Hello 消息的有效期。这个时间也是以秒为单位。如果一个路由器发送 Hello 消息，那么接受者会认为在一个 Holdtime 时间内，这个 Hello 消息是有效的。Holdtime 的值必须要比 Hellotime 的值大，而且至少是 Hellotime 值的 3 倍。

如果一个路由器上没有配置 Holdtime 值，则它会向由活动路由器发来的 Hello 消息学习到一个 Holdtime 值。如果 Hello 消息是被认证授权过的，则 Holdtime 值就只能通过学习来得到了。

同 Hellotime 一样，一个路由器必须引入那个在 Hello 消息中的 Holdtime 域所定义的 Holdtime 值。

一个状态为活动的路由器不能向其他路由器学习 Hellotime 和 Holdtime 值，尽管它也许会继续使用从前一任活动路由器那学到的 Hellotime 和 Holdtime 值。另外，它也许会使用手工配置的值。如果它没有学习到，而且也没有配置 Holdtime，则它会使用 10 秒作为默认值。

优先级：1 个字节，这个域用来选择活路由器和备份路由器。当把两个路由器的优先级进行比较时，优先级数值高的将获胜。如果两个路由器的优先级相同，则 IP 地址高的将获胜。

组：1 个字节，这个域定义了备份组。在令牌环网络中，它的值为 0 到 2，而在其他媒质中，它的值为 0 到 255 之间的数。

授权数据：8 字节，这个域包含了 8 个用作 password 的文本字符。如果授权数据没有被设置，则使用推荐的默认值：0x63 0x69 0x73 0x63 0x6F 0x00 0x00 0x00。

虚拟 IP 地址：4 字节，虚拟 IP 地址将在组中使用，如果一台路由器本身没有配置虚拟 IP 地址，那么它可以从活动路由器那发来的 Hello 消息中学到。而如果路由器没有设置虚拟 IP 地址，而且 Hello 消息已经被授权，则只能通过学习来获取这个地址。

（4）状态。

备份组中的每一台路由器都通过执行一个简单的状态机制来参与到这个协议中。下面我们就来描述一下我们在表面上所能看到的这个状态机制的一些运行情况。运行时可能会根据状态机制对不同功能的规定而在内部产生不同的操作过程。

1）Initial（初始状态）。

这是个开始的状态，它表明 HSRP 不在运行中。当配置改变或端口首次启动时就会进入这个状态。

2）Learn（学习状态）。

这是在路由器还没有确定虚拟 IP 地址，并且还没有收到一个从活动路由器发送来的已经认证过的 Hello 消息时的状态。在这个状态中，路由器仍然在等待着从活动路由器那里接收信息。

3）Listen（监听状态）。

路由器知道了虚拟 IP 地址，但它既不是活动路由器也不是备份路由器。并且该路由器是在从活动路由器或备份路由器那里监听 Hello 消息。

4）Speak（会话状态）。

路由器周期地发送 Hello 消息，并且积极地参与到活动路由器或备份路由器的选拔中。只有在它已经有了虚拟 IP 地址的前提下，它才能进入到这个状态。

5）Standby（备份状态）。

这个状态下的路由器作为下一个活动路由器的候选者，周期性地发送 Hello 消息。除了极短暂的情况外，每个组中最多只能有一个处于备份状态的路由器。

6）Active（激活状态）。

路由器的当前状态为把数据包转发到组的虚拟 MAC 地址。路由器周期地发送 Hello 消息。除了极短暂的情况外，每个组中最多只能有一个处于激活状态的路由器。

（5）时钟。

每台路由器都要维护 3 个时钟：一个激活时钟、一个备份时钟和一个 Hello 时钟。

激活时钟是用来监视活路由器的，在任何时候，只要路由器发现了从活动路由器发过来的被认证过的 Hello 消息，激活时钟就开始计时，直到到达 Hello 消息中所设定的 Holdtime 值为止。

备份时钟用于监视备份路由器。该时钟也是在路由器发现了从活动路由器发过来的被认证过的 Hello 消息后，随时开始计时，直到到达 Hello 消息中所设定的 Holdtime 值为止。

Hello 时钟是在每一个 Hellotime 时间段终止一次。如果路由器是处于会话、备份或激活状态，它会在 Hello 时钟停止时产生一个 Hello 消息。Hello 消息必须是不稳定的。

### 四、实验需要掌握的命令

switch(config-if)#no ip redirects
#禁止 ICMP 包的重定向

switch(config-if)#standby *group-number* priority *priority-number*
#设置在备份组 *group-number* 中，当前端口的优先级为 *priority-number*

switch(config-if)#standby *group-number* ip *ip-address*
#设置备份组 *group-number* 的虚拟地址为 *ip-address*

switch(config-if)# standy *group-number* preempt
#设置备份组 *group-number* 的模式为抢占模式

router(config-if)# standy *group-number* timers *hellotime holdtime*
#设置 Hello 时间和 Hold 时间参数

router(config-if)# standy *group-number* track *track-object* decrement *decrement-value*
#配置 HSRP 的上行端口的追踪

### 五、实验拓扑图与网络文件

（1）网络拓扑图。
网络拓扑图如图 12-1 所示。

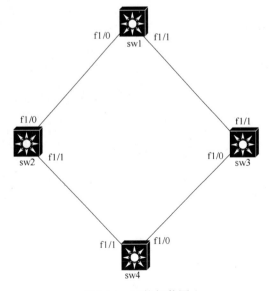

图 12-1　网络拓扑图

(2) IP 地址的规划。

IP 地址的规划如表 12-1 所示。

表 12-1 IP 地址规划

| 设备名称 | 端口号 | IP 地址 | 备注 |
|---|---|---|---|
| sw1 | vlan4 | 192.168.1.1/29 | 包含 f1/0、f1/1 端口 |
| sw2 | vlan2 | 192.168.2.2/24 | valn2 的 HSRP 组号为 1,虚拟地址为 192.168.2.1/29;valn3 的 HSRP 组号为 2,虚拟地址为 192.168.3.1/29;f1/1 配置为 trunk 模式;f1/0 配置为 Access 模式,属于 vlan4 |
| sw2 | vlan3 | 192.168.3.2/24 | |
| sw2 | vlan4 | 192.168.1.2/29 | |
| sw3 | vlan2 | 192.168.2.3/24 | valn2 的 HSRP 组号为 1,虚拟地址为 192.168.2.1/29;valn3 的 HSRP 组号为 2,虚拟地址为 192.168.3.1/29;f1/0 配置为 trunk 模式;f1/1 配置为 Access 模式,属于 vlan4 |
| sw3 | vlan3 | 192.168.3.3/24 | |
| sw3 | vlan4 | 192.168.1.3/29 | |
| sw4 | vlan2 | | vlan2 包含 f1/2-8;vlan2 包含 f1/9-15;f1/0 与 f1/1 配置为 trunk 模式 |
| sw4 | vlan3 | | |

(3) 网络文件。

```
#HSRP.net 文件开始
# HSRP lab
autostart = false
ghostios = true
sparsemem = true

[localhost]

    [[3660]]
    image = ..\c3660-is.bin
    ram = 160

    [[1720]]
    image = ..\C1700-Y.BIN
    ram = 16

    [[ROUTER sw1]]
    idlepc = 0x60719b98
    model = 3660
    slot1 = NM-16ESW
    f1/0 = sw2 f1/0
    f1/1 = sw3 f1/1

    [[router sw2]]
    idlepc = 0x60719b98
    model = 3660
    slot1 = NM-16ESW
    f1/1 = sw4 f1/1
    f1/2 = sw3 f1/2
```

```
[[router sw3]]
idlepc = 0x60719b98
model = 3660
slot1 = NM-16ESW
f1/0 = sw4 f1/0

[[router sw4]]
idlepc = 0x60719b98
model = 3660
slot1 = NM-16ESW
f1/2 = pc1 f0
f1/3 = pc2 f0

[[router pc1]]
model = 1720
idlepc = 0x8014e4ec

[[router pc2]]
model = 1720
idlepc = 0x8014e4ec
```
#HSRP.net 文件结束

## 六、实验配置步骤

（1）交换机 sw1 的配置步骤。

```
switch>enable
switch#vlan database
switch(vlan)#vlan 4        #创建 vlan4
switch(vlan)#exit
switch#configure terminal
switch(config)#hostname sw1
sw1(config-if-range)#interface range FastEthernet1/0 - 1
sw1(config-if-range)#switchport
sw1(config-if-range)#switchport access vlan 4
sw1(config-if-range)#speed 100
sw1(config-if-range)#duplex full
sw1(config-if-range)#no shutdown
sw1(config-if-range)#interface Vlan4
sw1(config-if)#ip address 192.168.1.1 255.255.255.248
sw1(config-if)#no shutdown
sw1(config-if)#exit
sw1(config)#ip route 192.168.2.0 255.255.254.0 Vlan4   #配置到网络 192.168.2.0/24 和 192.168.3.0/24 的路由
sw1(config)#do write
```

（2）交换机 sw2 的配置步骤。

```
switch>enable
switch#vlan database
switch(vlan)#vlan 2
switch(vlan)#vlan 3
```

```
switch(vlan)#vlan 4
switch(vlan)#exit
switch#configure terminal
switch(config)#hostname sw2
switch(config)#track 1 interface f1/0 line-protocol    #设置端口追踪
sw2(config)#interface FastEthernet1/0
sw2(config-if)#switchport
sw2(config-if)#switchport access vlan 4
sw2(config-if)#speed 100
sw2(config-if)#duplex full
sw2(config-if)#no shutdown
sw2(config-if)#interface FastEthernet1/1
sw2(config-if)#switchport
sw2(config-if)#switchport mode trunk
sw2(config-if)#switchport trunk allowed vlan all
sw2(config-if)#switchport trunk encapsulation dot1q
sw2(config-if)#speed 100
sw2(config-if)#duplex full
sw2(config-if)#no shutdown
sw2(config-if)#interface Vlan2
sw2(config-if)#ip address 192.168.2.2 255.255.255.0
sw2(config-if)#no ip redirects
#禁止 ICMP 包的重定向
sw2(config-if)#standby 1 ip 192.168.2.1
sw2(config-if)#standby 1 priority 110
sw2(config-if)# standby 1 preempt
sw2(config-if)#standby 1 track 1 decrement 20 #将上面的端口追踪应用到 hsrp 1 号组
#上面五条命令基于 vlan2 设置 HSRP 的配置
sw2(config-if)#no shutdown
sw2(config-if)#interface Vlan3
sw2(config-if)#ip address 192.168.3.2 255.255.255.0
sw2(config-if)#no ip redirects
#禁止 ICMP 包的重定向
sw2(config-if)# standby 2 ip 192.168.3.1
sw2(config-if)# standby 2 priority 105
sw2(config-if)# standby 2 preempt
sw2(config-if)# standby 2 authentication md5 key-string abcd
#上面五条命令基于 vlan3 设置 HSRP 的配置
sw2(config-if)#no shutdown
sw2(config-if)#interface Vlan4
sw2(config-if)#ip address 192.168.1.2 255.255.255.248
sw2(config-if)#no shutdown
sw2(config-if)#do write
```

（3）交换机 sw3 的配置步骤。

```
switch>enable
switch#vlan database
switch(vlan)#vlan 2
```

switch(vlan)#vlan 3
switch(vlan)#vlan 4
switch(vlan)#exit
switch#configure terminal
switch(config)#hostname sw3
sw3(config-if)#track 1 interface f1/1 line-protocol
sw3(config-if)#interface FastEthernet1/0
sw3(config-if)#switchport
sw3(config-if)#switchport mode trunk
sw3(config-if)#switchport trunk allowed vlan all
sw3(config-if)#switchport trunk encapsulation dot1q
sw3(config-if)#speed 100
sw3(config-if)#duplex full
sw3(config-if)#no shutdown
sw3(config-if)#interface FastEthernet1/1
sw3(config-if)#switchport
sw3(config-if)#switchport access vlan 4
sw3(config-if)#speed 100
sw3(config-if)#duplex full
sw3(config-if)#no shutdown
sw3(config-if)#interface Vlan2
sw3(config-if)#ip address 192.168.2.3 255.255.255.0
sw3(config-if)#no ip redirects
#禁止 ICMP 包的重定向
Sw3(config-if)#standby 1 ip 192.168.2.1
Sw3(config-if)#standby 1 priority 105
Sw3(config-if)# standby 1 preempt
#上面四条命令基于 vlan2 设置 HSRP 的配置
Sw3(config-if)#no shutdown
Sw3(config-if)#interface Vlan3
Sw3(config-if)#ip address 192.168.3.3 255.255.255.0
Sw3(config-if)#no ip redirects
#禁止 ICMP 包的重定向
Sw3(config-if)# standby 2 ip 192.168.3.1
Sw3(config-if)# standby 2 priority 110
Sw3(config-if)# standby 2 preempt
Sw3(config-if)# standby 2 authentication md5 key-string abcd
Sw3(config-if)#standby 2 track 1 decrement 20    #将上面的端口追踪应用到 HSRP 2 号组
#上面六条命令基于 vlan3 设置 HSRP 的配置
sw3(config-if)#no shutdown
sw3(config-if)#interface Vlan4
sw3(config-if)#ip address 192.168.1.3 255.255.255.248
sw3(config-if)#no shutdown
sw3(config-if)#do write

（4）交换机 sw4 的配置步骤。
switch>enable
switch#vlan database

```
switch(vlan)#vlan 2
switch(vlan)#vlan 3
switch(vlan)#exit
switch#configure terminal
switch(config)#hostname sw4
sw4(config)#interface range FastEthernet1/0 - 1
sw4(config-if-range)#switchport
sw4(config-if-range)#switchport mode trunk
sw4(config-if-range)#switchport trunk allowed vlan all
sw4(config-if-range)#switchport trunk encapsulation dot1q
sw4(config-if-range)#speed 100
sw4(config-if-range)#duplex full
sw4(config-if-range)#no shutdown
w4(config-if-range)#exit
w4(config)#no ip routing
#禁止路由，使此设备模拟一台二层交换机
w4(config)#write
```

## 七、实验测试和查看相关配置

（1）测试直连端口。

测试每一对直连端口通信是否正常，例如：在主机 sw1 上 ping 路由器 sw2 的 vlan4 端口，结果如下，说明这一对直连端口通信正常。如果没有收到任何 ICMP 的应答数据包，则说明这一对直连端口通信不正常，这时要查看端口的 IP 地址、子网掩码、speed、duplex 等参数的配置。

```
sw1>ping 192.168.1.2
Type escape sequence to abort
Sending 5, 100-byte ICMP Echos to 192.168.1.2, timeout is 2 seconds:
!!!!!
Success rate is 100 percent (5/5), round-trip min/avg/max = 152/202/260 ms
```

（2）查看 sw2 和 sw3 的 HSRP 配置。

```
sw2# show standby all
Vlan2 - Group 1    #vlan2 属于 HSRP 组 1
  State is Active   #在 HSRP 组 1 中，当前设备的 vlan2 为活动路由
    2 state changes, last state change 04:18:13
  Virtual IP address is 192.168.2.1    # HSRP 组 1 虚拟 IP 地址为 192.168.2.1
  Active virtual MAC address is 0000.0c07.ac01    # HSRP 组 1 虚拟 MAC 地址为 0000.0c07.ac01
    Local virtual MAC address is 0000.0c07.ac01 (v1 default)
  Hello time 3 sec, hold time 9 sec    #Hellotime 为 3 秒，Holdtime 为 9 秒
    Next hello sent in 2.584 secs    #下一个 hello 包发送的时间还有 2.584 秒
  Preemption enabled    # HSRP 组 1 的模式为抢占模式
  Active router is local    #当前设备的 vlan2 为活动路由
  Standby router is 192.168.2.3, priority 105 (expires in 6.276 sec)    #备份路由的 IP 地址为 192.168.2.3，优先级为 105
  Priority 110 (configured 110)    # HSRP 组 1 中，当前设备的优先级为 110
    Track object 1 state Up decrement 20    #追踪目标 1，追踪的端口状态为 up，如果为 down，则将当前设备中的 HSRP 组 1 优先级减 20
```

IP redundancy name is "hsrp-Vl2-1" (default)
Vlan3 - Group 2
　State is Standby
　　　13 state changes, last state change 03:07:57
　　Virtual IP address is 192.168.3.1
　　Active virtual MAC address is 0000.0c07.ac02
　　　Local virtual MAC address is 0000.0c07.ac02(v2 default)
　　Hello time 3 sec, hold time 9 sec
　　　Next hello sent in 2.480 secs
　　Authentication MD5, key-string "abcd"
　　Preemption enabled
　　Active router is 192.168.3.3, priority 110 (expires in 8.560 sec)
　　Standby router is local
　　Priority 105 (configured 105)
　　IP redundancy name is "hsrp-Vl3-2" (default)

router3#show standby all
Vlan2 - Group 1
　State is Standby
　　　16 state changes, last state change 03:06:44
　　Virtual IP address is 192.168.2.1
　　Active virtual MAC address is 0000.0c07.ac01
　　　Local virtual MAC address is 0000.0c07.ac01(v1 default)
　　Hello time 3 sec, hold time 9 sec
　　　Next hello sent in 0.128 secs
　　Preemption enabled
　　Active router is 192.168.2.2, priority 110 (expires in 8.396 sec)
　　Standby router is local
　　Priority 105 (configured 105)
　　IP redundancy name is "hsrp-Vl2-1" (default)
Vlan3 - Group 2
　State is Active
　　　1 state change, last state change 04:18:17
　　Virtual IP address is 192.168.3.1
　　Active virtual MAC address is 0000.0c07.ac02
　　　Local virtual MAC address is 0000.0c07.ac02 (v2 default)
　　Hello time 3 sec, hold time 9 sec
　　　Next hello sent in 0.132 secs
　　Authentication MD5, key-string "abcd"
　　Preemption enabled
　　Active router is local
　　Standby router is 192.168.3.2, priority 105 (expires in 7.728 sec)
　　Priority 110 (configured 110)
　　　Track object 1 state Up decrement 20
　　IP redundancy name is "hsrp-Vl3-2" (default)

（3）全网测试。

　　在sw4上接上一台主机，如果连接的端口属于vlan2，则IP地址一定是192.168.2.0/24网段的地址，如果连接的端口属于 vlan3，IP 地址一定是 192.168.3.0/24 网段的地址；然后在 PC 上 ping 192.168.1.1，看是否连通，如果不通，可能是路由的配置问题，也可能是端口的配置问题；如果是路由的问题，通过 show ip route 查看路由配置，如果路由没有问题，用 show interface *type number* 查看端口的配置，端口的配置要区分 Trunk 模式和 Access 模式，不能配置错误，否则也会导致网络不通。

## 八、注意问题

（1）端口的模式配置不能配置错误，是 Trunk 模式就不能配置为 Access 模式。

（2）端口的追踪只在 active 网关上设置，在 standby 网关上不用设置。

（3）本实验在配置冗余网关的同时也配置了网络负载均衡。

（4）在配置了 HSRP 的端口上必须用 no ip redirects 命令禁止 ICMP 包的重定向。

（5）虚拟 MAC 地址不能和相应的端口绑定，即在配置文件中不能出现与虚拟端口相关 mac-address-table static 0000.0c07.ac01 interface FastEthernet1/1 vlan 2 命令行，否则就无法 ping 通此网关。如果出现此命令行，可以用 no mac-address-table static 0000.0c07.ac01 将此命令行删除。其中 0000.0c07.ac01 为 VRRP1 组默认虚拟 MAC 地址。

# 实验十三  交换机综合实验

本实验模拟一个中小型企业的网络，使用核心层、汇聚层、接入层三层的结构，综合应用前面所学的知识进行综合训练，从而将前面的知识点串联起来，使学生掌握知识点的综合应用，从而提高学生的实际处理问题的能力。

## 一、实验目的

（1）网络 IP 地址的规划；
（2）交换机远程登录、VLAN、Trunk、管理 IP、端口 IP、VTP、静态路由、NAT 的配置；
（3）网络的测试与调试。

要求：网络 IP 地址规划；Web 服务器是公网地址，不管局域网还是广域网都可以直接访问；将汇聚层交换机和接入层的 vlan 通过 VTP 进行管理，同时将每个 vlan 分配多个端口；接入层交换机与汇聚层交换机之间可以进行多个 vlan 通信；汇聚层交换机与核心层交换机通过路由进行通信；NAT 路由器上配置 NAT，使局域网的所有主机都可以访问广域网；保证所有主机之间都可以互相通信。

## 二、实验拓扑图和网络文件

由于此实验的设备较多，用一台计算机无法模拟，所以此实验应用 DYNAMIPS 的分布式网络环境进行模拟，下面的拓扑文件就是应用两台计算机进行模拟的。如果有条件的话，此实验最好用三台计算机进行模拟，这样速度更快，在做测试时也不容易出现丢包问题。

（1）网络拓扑图。
网络拓扑图如图 13-1 所示。
（2）网络文件。

```
autostart = false
ghostios = true
sparsemem = true

[192.168.7.158:7200]    #本地主机的 IP 地址和端口号

    [[3660]]
    image = ..\C3660-IS.BIN
    ram = 160

    [[3640]]
    image = ..\C3640-IK.BIN
    ram = 160

    #[[1720]]
    #image = ..\C1700-Y.BIN
```

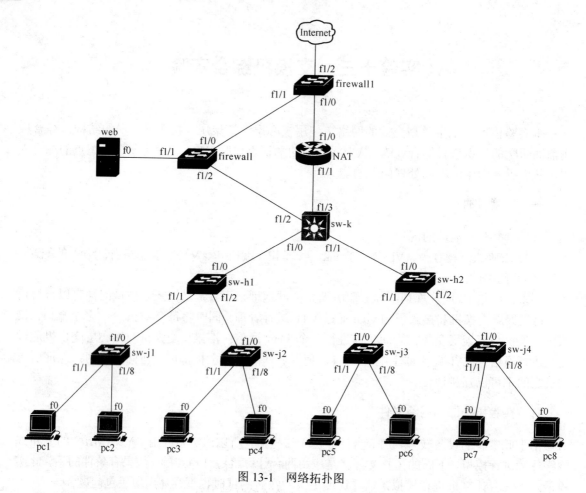

图 13-1 网络拓扑图

#ram = 16

[[ROUTER firewall1]]
model = 3660
idlepc = 0x60719b98
slot1 = NM-16ESW
f1/0 = NAT f1/0
f1/1 = firewall f1/0
f1/2 = NIO_gen_eth:\Device\NPF_{E2B31736-C716-4215-A9CB-ADA78C760F52}

[[ROUTER firewall]]
model = 3660
idlepc = 0x60719b98
slot1 = NM-16ESW
f1/1 = web f0
f1/2 = sw-k f1/2

[[ROUTER NAT]]
model = 3660
idlepc = 0x60719b98

slot1 = NM-16ESW
f1/1 = sw-k f1/3

[[ROUTER sw-k]]
model = 3660
idlepc = 0x60719b98
slot1 = NM-16ESW
f1/0 = sw-h1 f1/0
f1/1 = sw-h2 f1/0

[[router sw-h1]]
model = 3660
idlepc = 0x6064b000
slot1 = NM-16ESW
f1/1 = sw-j1 f1/0
f1/2 = sw-j2 f1/0

[[router sw-h2]]
model = 3660
idlepc = 0x6064b000
slot1 = NM-16ESW
f1/1 = sw-j3 f1/0
f1/2 = sw-j4 f1/0

[[router sw-j1]]
model = 3640
idlepc = 0x6050c860
slot1 = NM-16ESW
f1/1 = pc1 f0
f1/8 = pc2 f0

[[router sw-j2]]
model = 3640
idlepc = 0x6050c860
slot1 = NM-16ESW
f1/1 = pc3 f0
f1/8 = pc4 f0

[192.168.7.157:7200]  #远程 Dynamips 服务器的 IP 地址与端口号

workingdir = e:\zonghe  #远程 Dynamips 服务器的工作目录

[[3640]]
image = c:\program files\dynamips\images\C3640-IK.BIN
ram = 160

[[1720]]
image = c:\program files\dynamips\images\C1700-Y.BIN

ram = 16

[[router sw-j3]]
model = 3640
idlepc = 0x6050c860
slot1 = NM-16ESW
f1/1 = pc5 f0
f1/8 = pc6 f0

[[router sw-j4]]
model = 3640
idlepc = 0x6050c860
slot1 = NM-16ESW
f1/1 = pc7 f0
f1/8 = pc8 f0

[[router pc1]]
model = 1720
idlepc = 0x8014e4ec

[[router pc2]]
model = 1720
idlepc = 0x8014e4ec

[[router pc3]]
model = 1720
idlepc = 0x8014e4ec

[[router pc4]]
model = 1720
idlepc = 0x8014e4ec

[[router pc5]]
model = 1720
idlepc = 0x8014e4ec

[[router pc6]]
model = 1720
idlepc = 0x8014e4ec

[[router pc7]]
model = 1720
idlepc = 0x8014e4ec

[[router pc8]]
model = 1720
idlepc = 0x8014e4ec

```
[[router web]]
    model = 1720
idlepc = 0x8014e4ec
```

## 三、IP 地址的规划

地址分配的几个基本规则如下：

**规则一：体系化编址。**

体系化其实就是结构化、组织化，以企业的具体需求和组织结构为原则对整个网络地址进行有条理的规划。一般这个规划的过程是由大局、整体着眼，然后逐级由大到小分割、划分的。这其实跟实际的物理地址分配原则是一样的，肯定是先划分省市、再细分出县区、道路、街巷，最后是门牌。从网络总体来说，对体系化编制，由于相邻或者具有相同服务性质的主机或办公群落都在 IP 地址上且是连续的，这样在各个区块的边界路由设备上便于进行有效的路由汇总，使整个网络的结构清晰，路由信息明确，也能减小路由器中的路由表。而每个区域的地址与其他的区域地址相对独立，也便于独立地灵活管理。

**规则二：可持续扩展性。**

可持续扩展性其实就是在初期规划时为将来的网络拓展考虑，眼光要放得长远一些，在将来很可能增大规模的区块中要留出较大的余地。IP 地址最开始是按有类划分的，A、B、C 各类标准网段都只能严格按照规定使用地址。但现在发展到了无类阶段，由于可以自由规划子网的大小和实际的主机数，所以使得地址资源分配得更加合理，无形中就增大了网络的可拓展性。虽然在网络初期的一段可能很长的时间里，未合理考虑余量的 IP 地址规划也能满足需要，但是当一个局部区域出现高增长，或者整体的网络规模不断增大，这时不合理的规划很可能必须重新部署局部甚至整体的 IP 地址，这在一个中、大型网络中就绝不是一个轻松的工作了。

**规则三：按需分配公网 IP。**

相对于私有 IP 而言，公网 IP 是不能由自己完全做主要求的，而是 ISP 等机构统一分配和租用的。这就造成了公网 IP 要稀缺得多，所以对公网 IP 必须按实际需求来分配。如对外提供服务的服务器群组区域，不仅要够用，还得预留出余量；而员工部门等仅需要浏览 Internet 等基本需求的区域，可以通过 NAT（网络地址转换）来使多个节点共享一个或几个公网 IP；最后，那些只对内部提供服务，或只限于内部通信的主机自然不用分配公网 IP 了。公网 IP 具体的分配必须根据实际的需求，进行合理的规划。

本实验的 IP 地址规划表如表 13-1 所示。

表 13-1  IP 地址规划

| 设备名称 | 端口号 | IP 地址 | 备注 |
| --- | --- | --- | --- |
| firewall | f1/0 | 192.168.1.17/30 | 为了节省公网地址（用 192.168.1.0/24 内网地址代替公网地址） |
| | f1/1 | 10.1.2.17/28 | |
| | f1/2 | 192.168.1.2/30 | |
| firewall1 | f1/0 | 10.1.2.6/30 | |
| | f1/1 | 192.168.1.18/30 | 为了节省公网地址 |
| | f1/2 | 10.1.2.9/30 | |

续表

| 设备名称 | 端口号 | IP 地址 | 备注 |
|---|---|---|---|
| NAT | f1/0 | 10.1.2.5/30 | |
| | f1/1 | 192.168.1.6/30 | |
| sw-k | f1/0 | 192.168.1.9/30 | |
| | f1/1 | 192.168.1.13/30 | |
| | f1/2 | 192.168.1.1/30 | |
| | f1/3 | 192.168.1.5/30 | |
| sw-h1 | f1/0 | 192.168.1.10/30 | 与核心交换机 sw-k 端口 f1/0 通过路由方式连接 |
| | f1/1 | | 与接入层交换机 sw-j1 端口 f1/0 连接，配置 trunk |
| | f1/2 | | 与接入层交换机 sw-j2 端口 f1/0 连接，配置 trunk |
| | vlan 4 | 192.168.4.1/24 | 作为 192.168.4.1/24 网段的网关 |
| | vlan 5 | 192.168.5.1/24 | 作为 192.168.5.1/24 网段的网关 |
| | vlan 6 | 192.168.6.1/24 | 作为 192.168.6.1/24 网段的网关 |
| | vlan 7 | 192.168.7.1/24 | 作为 192.168.7.1/24 网段的网关 |
| sw-h2 | f1/0 | 192.168.1.14/30 | 与核心交换机 sw-k 端口 f1/1 通过路由方式连接 |
| | f1/1 | | 与接入层交换机 sw-j3 端口 f1/0 连接，配置 trunk |
| | f1/2 | | 与接入层交换机 sw-j4 端口 f1/0 连接，配置 trunk |
| | vlan 8 | 192.168.8.1/24 | 作为 192.168.8.1/24 网段的网关 |
| | vlan 9 | 192.168.9.1/24 | 作为 192.168.9.1/24 网段的网关 |
| | vlan 10 | 192.168.10.1/24 | 作为 192.168.10.1/24 网段的网关 |
| | vlan 11 | 192.168.11.1/24 | 作为 192.168.11.1/24 网段的网关 |
| sw-j1 | f1/0 | | |
| | f1/1-7 | | 端口 f1/1-7 分配给 vlan 4 |
| | f1/8-15 | | 端口 f1/8-15 分配给 vlan 5 |
| sw-j2 | f1/0 | | |
| | f1/1-7 | | 端口 f1/1-7 分配给 vlan 6 |
| | f1/8-15 | | 端口 f1/8-15 分配给 vlan 7 |
| sw-j3 | f1/0 | | |
| | f1/1-7 | | 端口 f1/1-7 分配给 vlan 8 |
| | f1/8-15 | | 端口 f1/8-15 分配给 vlan 9 |
| sw-j4 | f1/0 | | |
| | f1/1-7 | | 端口 f1/1-7 分配给 vlan 10 |
| | f1/8-15 | | 端口 f1/8-15 分配给 vlan 11 |
| pc1 | f0 | 192.168.4.2/24 | 与接入层交换机 sw-j1 端口 f1/1 连接 |
| pc2 | f0 | 192.168.5.2/24 | 与接入层交换机 sw-j1 端口 f1/8 连接 |
| pc3 | f0 | 192.168.6.2/24 | 与接入层交换机 sw-j2 端口 f1/1 连接 |

续表

| 设备名称 | 端口号 | IP 地址 | 备注 |
|---|---|---|---|
| pc4 | f0 | 192.168.7.2/24 | 与接入层交换机 sw-j2 端口 f1/8 连接 |
| pc5 | f0 | 192.168.8.2/24 | 与接入层交换机 sw-j3 端口 f1/1 连接 |
| pc6 | f0 | 192.168.9.2/24 | 与接入层交换机 sw-j3 端口 f1/8 连接 |
| pc7 | f0 | 192.168.10.2/24 | 与接入层交换机 sw-j4 端口 f1/1 连接 |
| pc8 | f0 | 192.168.11.2/24 | 与接入层交换机 sw-j4 端口 f1/8 连接 |
| web | f0 | 10.1.2.18/28 | 与 firewall 端口 f1/1 连接，作为 DMZ 区服务器 |

四、实验配置步骤

（1）NAT 交换机的配置步骤。

switch>enable
switch #configure terminal
switch (config)#hostname NAT
NAT (config)#interface FastEthernet 1/0
NAT (config-if)#no switchport
NAT (config-if)#ip address 10.1.2.5 255.255.255.252
NAT (config-if)#speed 100
NAT (config-if)#duplex full
NAT (config-if)#ip nat outside
NAT (config-if)#no shutdown
NAT (config-if)#interface FastEthernet 1/1
NAT (config-if)#no switchport
NAT (config-if)#ip address 192.168.1.6 255.255.255.252
NAT (config-if)#speed 100
NAT (config-if)#duplex full
NAT (config-if)#ip nat inside
NAT (config-if)#no shutdown
NAT (config-if)#exit
NAT (config)# ip route 0.0.0.0 0.0.0.0 f1/0    #默认路由。
NAT (config)# ip route 192.168.0.0 255.255.0.0 f1/1    #此路由为所有内网用户访问外网的数据包提供返回路径
NAT (config)# ip nat inside source list 1 interface FastEthernet1/0 overload
NAT (config)# access-list 1 permit 192.168.0.0 0.0.255.255
NAT (config)#exit
NAT #write

（2）firewall 交换机（模拟防火墙）的配置步骤。

switch>enable
switch #configure terminal
switch (config)#hostname firewall
firewall (config)# interface FastEthernet1/0
firewall (config-if)# no switchport
firewall (config-if)# ip address 192.168.1.17 255.255.255.252
firewall (config-if)# duplex full

```
firewall (config-if)# speed 100
firewall (config-if)#no shutdown
firewall (config-if)# interface FastEthernet1/1
firewall (config-if)# no switchport
firewall (config-if)# ip address 10.1.2.17 255.255.255.240
firewall (config-if)# duplex full
firewall (config-if)# speed 100
firewall (config-if)#no shutdown
firewall (config-if)# interface FastEthernet1/2
firewall (config-if)# no switchport
firewall (config-if)# ip address 192.168.1.2 255.255.255.252
firewall (config-if)# duplex full
firewall (config-if)# speed 100
firewall (config-if)#no shutdown
firewall (config-if)#exit
firewall (config)# ip route 0.0.0.0 0.0.0.0 FastEthernet1/0    #默认路由
firewall (config)# ip route 192.168.0.0 255.255.0.0 FastEthernet1/2    #此路由为所有内网用户访问 web 服务器所在网段的数据包提供返回路径
firewall (config)#exit
firewall #write
```

（3）firewall1 交换机（模拟防火墙）的配置步骤。

```
switch>enable
switch #configure terminal
switch (config)#hostname firewall1
firewall1 (config)# interface FastEthernet1/0
firewall1 (config-if)# no switchport
firewall1 (config-if)# ip address 10.1.2.6 255.255.255.252
firewall1 (config-if)# duplex full
firewall1 (config-if)# speed 100
firewall1 (config-if)#no shutdown
firewall1 (config-if)# interface FastEthernet1/1
firewall1 (config-if)# no switchport
firewall1 (config-if)# ip address 192.168.1.18 255.255.255.252
firewall1 (config-if)# duplex full
firewall1 (config-if)# speed 100
firewall1 (config-if)#no shutdown
firewall1 (config-if)# interface FastEthernet1/2
firewall1 (config-if)# no switchport
firewall1 (config-if)# ip address 10.1.2.9 255.255.255.252
firewall1 (config-if)# duplex full
firewall1 (config-if)# speed 100
firewall1 (config-if)#no shutdown
firewall1 (config-if)#exit
firewall1 (config)# ip route 0.0.0.0 0.0.0.0 FastEthernet1/2    #默认路由
firewall1 (config)# ip route 10.1.2.16 255.255.255.240 FastEthernet1/1    #此路由为外网用户访问 web 服务器所在网段的数据包提供路径
firewall1 (config)#exit
```

firewall1 #write

(4) sw-k 核心层交换机的配置步骤。

switch>enable
switch #configure terminal
switch (config)#hostname sw-k
sw-k (config)# interface FastEthernet1/0
sw-k (config-if)# no switchport
sw-k (config-if)# ip address 192.168.1.9 255.255.255.252
sw-k (config-if)# duplex full
sw-k (config-if)# speed 100
sw-k (config-if)#no shutdown
sw-k (config-if)# interface FastEthernet1/1
sw-k (config-if)# no switchport
sw-k (config-if)# ip address 192.168.1.13 255.255.255.252
sw-k (config-if)# duplex full
sw-k (config-if)# speed 100
sw-k (config-if)#no shutdown
sw-k (config)# interface FastEthernet1/2
sw-k (config-if)# no switchport
sw-k (config-if)# ip address 192.168.1.1 255.255.255.252
sw-k (config-if)# duplex full
sw-k (config-if)# speed 100
sw-k (config-if)#no shutdown
sw-k (config-if)# interface FastEthernet1/3
sw-k (config-if)# no switchport
sw-k (config-if)# ip address 192.168.1.5 255.255.255.252
sw-k (config-if)# duplex full
sw-k (config-if)# speed 100
sw-k (config-if)#no shutdown
sw-k (config-if)#exit
sw-k (config)# ip route 0.0.0.0 0.0.0.0 FastEthernet1/3    #默认路由
sw-k (config)# ip route 10.1.2.16 255.255.255.240 FastEthernet1/2   #此路由为外网用户访问 web 服务器所在网段的数据包提供路径
sw-k (config)# ip route 192.168.4.0 255.255.252.0 FastEthernet1/0   #此路由为目的地址为 192.168.4.0/22 网段用户的数据包提供路径
sw-k (config)# ip route 192.168.8.0 255.255.252.0 FastEthernet1/1   #此路由为目的地址为 192.168.8.0/22 网段用户的数据包提供路径
sw-k (config)#exit
sw-k #write

(5) sw-h1 汇聚层交换机的配置步骤。

switch>enable
switch #configure terminal
switch (config)#hostname sw-h1
sw-h1 (config)#exit
sw-h1 #vlan database
sw-h1 (vlan)#vtp server
sw-h1 (vlan)#vtp domain sw-h1

```
sw-h1 (vlan)#vtp password 123456
sw-h1 (vlan)#vtp pruning
sw-h1 (vlan)#vtp v2-mode
sw-h1 (vlan)#valn 4
sw-h1 (vlan)#valn 5
sw-h1 (vlan)#valn 6
sw-h1 (vlan)#valn 7
sw-h1 (vlan)#apply
sw-h1 (vlan)#exit
sw-h1 #configure terminal
sw-h1 (config)# interface FastEthernet1/0
sw-h1 (config-if)# no switchport
sw-h1 (config-if)# ip address 192.168.1.10 255.255.255.252
sw-h1 (config-if)# duplex full
sw-h1 (config-if)# speed 100
sw-h1 (config-if)#no shutdown
sw-h1 (config-if)# interface FastEthernet1/1
sw-h1 (config-if)# switchport
sw-h1 (config-if)# switchport mode trunk
sw-h1 (config-if)# switchport trunk encapsulation dot1q
sw-h1 (config-if)# switchport trunk allowed vlan all
sw-h1 (config-if)# duplex full
sw-h1 (config-if)# speed 100
sw-h1 (config-if)#no shutdown
sw-h1 (config-if)# interface FastEthernet1/2
sw-h1 (config-if)# switchport
sw-h1 (config-if)# switchport mode trunk
sw-h1 (config-if)# switchport trunk encapsulation dot1q
sw-h1 (config-if)# switchport trunk allowed vlan all
sw-h1 (config-if)# duplex full
sw-h1 (config-if)# speed 100
sw-h1 (config-if)#no shutdown
sw-h1 (config-if)# interface Vlan 4
sw-h1 (config-if)#ip address 192.168.4.1 255.255.255.0
sw-h1 (config-if)#no shutdown
sw-h1 (config-if)#interface Vlan 5
sw-h1 (config-if)#ip address 192.168.5.1 255.255.255.0
sw-h1 (config-if)#no shutdown
sw-h1 (config-if)#interface Vlan 6
sw-h1 (config-if)#ip address 192.168.6.1 255.255.255.0
sw-h1 (config-if)#no shutdown
sw-h1 (config-if)#interface Vlan 7
sw-h1 (config-if)#ip address 192.168.7.1 255.255.255.0
sw-h1 (config-if)#no shutdown
sw-h1 (config-if)#exit
sw-h1 (config)# ip route 0.0.0.0 0.0.0.0 FastEthernet1/0    #默认路由
sw-h1 (config)#exit
sw-h1 #write
```

（6）sw-h2 汇聚层交换机的配置步骤。
switch>enable
switch #configure terminal
switch (config)#hostname sw-h2
sw-h2 (config)#exit
sw-h2 #vlan database
sw-h2 (vlan)#vtp server
sw-h2 (vlan)#vtp domain sw-h2
sw-h2 (vlan)#vtp password 123321
sw-h2 (vlan)#vtp pruning
sw-h2 (vlan)#vtp v2-mode
sw-h2 (vlan)#valn 8
sw-h2 (vlan)#valn 9
sw-h2 (vlan)#valn 10
sw-h2 (vlan)#valn 11
sw-h2 (vlan)#apply
sw-h2 (vlan)#exit
sw-h2 #configure terminal
sw-h2 (config)# interface FastEthernet1/0
sw-h2 (config-if)# no switchport
sw-h2 (config-if)# ip address 192.168.1.14 255.255.255.252
sw-h2 (config-if)# duplex full
sw-h2 (config-if)# speed 100
sw-h2 (config-if)#no shutdown
sw-h2 (config-if)# interface FastEthernet1/1
sw-h2 (config-if)# switchport
sw-h2 (config-if)# switchport mode trunk
sw-h2 (config-if)# switchport trunk encapsulation dot1q
sw-h2 (config-if)# switchport trunk allowed vlan all
sw-h2 (config-if)# duplex full
sw-h2 (config-if)# speed 100
sw-h2 (config-if)#no shutdown
sw-h2 (config-if)# interface FastEthernet1/2
sw-h2 (config-if)# switchport
sw-h2 (config-if)# switchport mode trunk
sw-h2 (config-if)# switchport trunk encapsulation dot1q
sw-h2 (config-if)# switchport trunk allowed vlan all
sw-h2 (config-if)# duplex full
sw-h2 (config-if)# speed 100
sw-h2 (config-if)#no shutdown
sw-h2 (config-if)# interface Vlan 8
sw-h2 (config-if)# ip address 192.168.8.1 255.255.255.0
sw-h2 (config-if)#no shutdown
sw-h2 (config-if)#interface Vlan 9
sw-h2 (config-if)# ip address 192.168.9.1 255.255.255.0
sw-h2 (config-if)#no shutdown
sw-h2 (config-if)#interface Vlan 10

sw-h2 (config-if)#ip address 192.168.10.1 255.255.255.0
sw-h2 (config-if)#no shutdown
sw-h2 (config-if)#interface Vlan 11
sw-h2 (config-if)#ip address 192.168.11.1 255.255.255.0
sw-h2 (config-if)#no shutdown
sw-h2 (config-if)#exit
sw-h2 (config)# ip route 0.0.0.0 0.0.0.0 FastEthernet1/0    #默认路由
sw-h2 (config)#exit
sw-h2 #write

（7）sw-j1 接入层交换机的配置步骤。

switch>enable
switch #configure terminal
switch (config)#hostname sw-j1
sw-j1 (config)#exit
sw-j1 #vlan database
sw-j1 (vlan)#vtp domain sw-h1
sw-j1 (vlan)#vtp client
sw-j1 (vlan)#vtp password 123456
sw-j1 (vlan)#vtp v2-mode
sw-j1 (vlan)#exit
sw-j1 #configure terminal
sw-j1 (config)# interface FastEthernet1/0
sw-j1 (config-if)# switchport mode trunk
sw-j1 (config-if)# switchport trunk encapsulation dot1q
sw-j1 (config-if)# switchport trunk allowed vlan all
sw-j1 (config-if)# duplex full
sw-j1 (config-if)# speed 100
sw-j1 (config-if)#no shutdown
sw-j1 (config-if)# interface range FastEthernet1/1 - 7
sw-j1 (config-if-range)# switchport mode access
sw-j1 (config-if-range)# switchport access vlan 4
sw-j1 (config-if-range)# duplex full
sw-j1 (config-if-range)# speed 100
sw-j1 (config-if-range)#no shutdown
sw-j1 (config-if-range)# interface range FastEthernet1/8 - 15
sw-j1 (config-if-range)# switchport mode access
sw-j1 (config-if-range)# switchport access vlan 5
sw-j1 (config-if-range)# duplex full
sw-j1 (config-if-range)# speed 100
sw-j1 (config-if-range)#no shutdown
sw-j1 (config-if)#interface Vlan 4
sw-j1 (config-if)#ip address 192.168.4.254 255.255.255.0   #交换机 sw-j1 管理 IP 地址
sw-j1 (config-if)#no shutdown
sw-j1 (config-if)#exit
sw-j1 (config)#no ip routing
sw-j1 (config)#ip default-gateway 192.168.4.1   #设置交换机的默认网关
sw-j1 (config)#end

sw-j1 #write

(8) sw-j2 接入层交换机的配置步骤。

switch>enable
switch#configure terminal
switch(config)#hostname sw-j2
sw-j2(config)#exit
sw-j2#vlan database
sw-j2(vlan)#vtp domain sw-h1
sw-j2(vlan)#vtp client
sw-j2(vlan)#vtp password 123456
sw-j2(vlan)#vtp v2-mode
sw-j2(vlan)#exit
sw-j2#configure terminal
sw-j2(config)# interface FastEthernet1/0
sw-j2(config-if)# switchport mode trunk
sw-j2(config-if)# switchport trunk encapsulation dot1q
sw-j2(config-if)# switchport trunk allowed vlan all
sw-j2(config-if)# duplex full
sw-j2(config-if)# speed 100
sw-j2(config-if)#no shutdown
sw-j2(config-if)# interface range FastEthernet1/1 - 7
sw-j2(config-if-range)# switchport mode access
sw-j2(config-if-range)# switchport access vlan 6
sw-j2(config-if-range)# duplex full
sw-j2(config-if-range)# speed 100
sw-j2(config-if-range)#no shutdown
sw-j2(config-if-range)# interface range FastEthernet1/8 - 15
sw-j2(config-if-range)# switchport mode access
sw-j2(config-if-range)# switchport access vlan 7
sw-j2(config-if-range)# duplex full
sw-j2(config-if-range)# speed 100
sw-j2(config-if-range)#no shutdown
sw-j2(config-if)#interface Vlan 6
sw-j2(config-if)#ip address 192.168.6.254 255.255.255.0    #交换机 sw-j2 管理 IP 地址
sw-j2(config-if)#no shutdown
sw-j2(config-if)#exit
sw-j2(config)#no ip routing
sw-j2(config)#ip default-gateway 192.168.6.1    #设置交换机的默认网关
sw-j2(config)#end
sw-j2(config-if)#end
sw-j2#write

(9) sw-j3 接入层交换机的配置步骤。

switch>enable
switch#configure terminal
switch(config)#hostname sw-j3
sw-j3(config)#exit
sw-j3#vlan database

```
sw-j3(vlan)#vtp domain sw-h2
sw-j3(vlan)#vtp client
sw-j3(vlan)#vtp password 123321
sw-j3(vlan)#vtp v2-mode
sw-j3(vlan)#exit
sw-j3#configure terminal
sw-j3(config)# interface FastEthernet1/0
sw-j3(config-if)# switchport mode trunk
sw-j3(config-if)# switchport trunk encapsulation dot1q
sw-j3(config-if)# switchport trunk allowed vlan all
sw-j3(config-if)# duplex full
sw-j3(config-if)# speed 100
sw-j3(config-if)#no shutdown
sw-j3(config-if)# interface range FastEthernet1/1 - 7
sw-j3(config-if-range)# switchport mode access
sw-j3(config-if-range)# switchport access vlan 8
sw-j3(config-if-range)# duplex full
sw-j3(config-if-range)# speed 100
sw-j3(config-if-range)#no shutdown
sw-j3(config-if-range)# interface range FastEthernet1/8 - 15
sw-j3(config-if-range)# switchport mode access
sw-j3(config-if-range)# switchport access vlan 9
sw-j3(config-if-range)# duplex full
sw-j3(config-if-range)# speed 100
sw-j3(config-if-range)#no shutdown
sw-j3(config-if)#interface Vlan 8
sw-j3(config-if)#ip address 192.168.8.254 255.255.255.0    #交换机 sw-j3 管理 IP 地址
sw-j3(config-if)#no shutdown
sw-j3(config-if)#exit
sw-j3(config)#no ip routing
sw-j3(config)#ip default-gateway 192.168.8.1    #设置交换机的默认网关
sw-j3(config)#end
sw-j3(config-if)#end
sw-j3#write
```

（10）sw-j4 接入层交换机的配置步骤。

```
switch>enable
switch#configure terminal
switch(config)#hostname sw-j4
sw-j4(config)#exit
sw-j4#vlan database
sw-j4(vlan)#vtp domain sw-h2
sw-j4(vlan)#vtp client
sw-j4(vlan)#vtp password 123321
sw-j4(vlan)#vtp v2-mode
sw-j4(vlan)#exit
sw-j4#configure terminal
sw-j4(config)# interface FastEthernet1/0
```

sw-j4(config-if)# switchport mode trunk
sw-j4(config-if)# switchport trunk encapsulation dot1q
sw-j4(config-if)# switchport trunk allowed vlan all
sw-j4(config-if)# duplex full
sw-j4(config-if)# speed 100
sw-j4(config-if)#no shutdown
sw-j4(config-if)# interface range FastEthernet1/1 - 7
sw-j4(config-if-range)# switchport mode access
sw-j4(config-if-range)# switchport access vlan 10
sw-j4(config-if-range)# duplex full
sw-j4(config-if-range)# speed 100
sw-j4(config-if-range)#no shutdown
sw-j4(config-if-range)# interface range FastEthernet1/8 - 15
sw-j4(config-if-range)# switchport mode access
sw-j4(config-if-range)# switchport access vlan 11
sw-j4(config-if-range)# duplex full
sw-j4(config-if-range)# speed 100
sw-j4(config-if-range)#no shutdown
sw-j4(config-if)#interface Vlan 10
sw-j4(config-if)#ip address 192.168.10.254 255.255.255.0    #交换机 sw-j4 管理 IP 地址
sw-j4(config-if)#no shutdown
sw-j4(config-if)#exit
sw-j4(config)#no ip routing
sw-j4(config)#ip default-gateway 192.168.10.1    #设置交换机的默认网关
sw-j4(config)#end
sw-j4(config-if)#end
sw-j4#write

（11）主机 pc1 的配置步骤。

switch>enable
switch #configure terminal
switch (config)#hostname pc1
pc1 (config)#interface FastEthernet0
pc1 (config-if)#ip address 192.168.4.2 255.255.255.0
pc1 (config-if)#duplex full
pc1 (config-if)#no shutdown
pc1 (config-if)#exit
pc1 (config)#no ip routing
pc1 (config)#ip default-gateway 192.168.4.1
pc1 (config)#exit
pc1#write

（12）主机 pc2 的配置步骤。

switch>enable
switch #configure terminal
switch (config)#hostname pc2
pc2 (config)#interface FastEthernet0
pc2 (config-if)#ip address 192.168.5.2 255.255.255.0
pc2 (config-if)#duplex full

pc2 (config-if)#no shutdown
pc2 (config-if)#exit
pc2 (config)#no ip routing
pc2 (config)#ip default-gateway 192.168.5.1
pc2 (config)#exit
pc2#write

（13）主机 pc3 的配置步骤。

switch>enable
switch #configure terminal
switch (config)#hostname pc3
pc31 (config)#interface FastEthernet0
pc3 (config-if)#ip address 192.168.6.2 255.255.255.0
pc3 (config-if)#duplex full
pc3 (config-if)#no shutdown
pc3 (config-if)#exit
pc3 (config)#no ip routing
pc3 (config)#ip default-gateway 192.168.6.1
pc3 (config)#exit
pc3#write

（14）主机 pc4 的配置步骤。

switch>enable
switch #configure terminal
switch (config)#hostname pc4
pc4 (config)#interface FastEthernet0
pc4 (config-if)#ip address 192.168.7.2 255.255.255.0
pc4 (config-if)#duplex full
pc4 (config-if)#no shutdown
pc4(config-if)#exit
pc4 (config)#no ip routing
pc4 (config)#ip default-gateway 192.168.7.1
pc4 (config)#exit
pc4#write

（15）主机 pc5 的配置步骤。

switch>enable
switch #configure terminal
switch (config)#hostname pc5
pc5 (config)#interface FastEthernet0
pc5 (config-if)#ip address 192.168.8.2 255.255.255.0
pc5 (config-if)#duplex full
pc5 (config-if)#no shutdown
pc5 (config-if)#exit
pc5 (config)#no ip routing
pc5 (config)#ip default-gateway 192.168.8.1
pc5 (config)#exit
pc5#write

（16）主机 pc6 的配置步骤。

switch>enable

switch #configure terminal
switch (config)#hostname pc6
pc6 (config)#interface FastEthernet0
pc6 (config-if)#ip address 192.168.9.2 255.255.255.0
pc6 (config-if)#duplex full
pc6 (config-if)#no shutdown
pc6 (config-if)#exit
pc6 (config)#no ip routing
pc6 (config)#ip default-gateway 192.168.9.1
pc6 (config)#exit
pc6#write

（17）主机 pc7 的配置步骤。

switch>enable
switch #configure terminal
switch (config)#hostname pc7
pc7 (config)#interface FastEthernet0
pc7 (config-if)#ip address 192.168.10.2 255.255.255.0
pc7 (config-if)#duplex full
pc7 (config-if)#no shutdown
pc7 (config-if)#exit
pc7 (config)#no ip routing
pc7 (config)#ip default-gateway 192.168.10.1
pc71 (config)#exit
pc7#write

（18）主机 pc8 的配置步骤。

switch>enable
switch #configure terminal
switch (config)#hostname pc8
pc8 (config)#interface FastEthernet0
pc8 (config-if)#ip address 192.168.11.2 255.255.255.0
pc8 (config-if)#duplex full
pc8 (config-if)#no shutdown
pc8 (config-if)#exit
pc8 (config)#no ip routing
pc8 (config)#ip default-gateway 192.168.11.1
pc8 (config)#exit
pc8#write

（19）主机 web 的配置步骤。

switch>enable
switch #configure terminal
switch (config)#hostname web
web (config)#interface FastEthernet0
web (config-if)#ip address 10.1.2.18 255.255.255.240
web (config-if)#duplex full
web (config-if)#no shutdown
web (config-if)#exit
web (config)#no ip routing

```
web (config)#ip default-gateway 10.1.2.17
web (config)#exit
web #write
```

### 五、实验测试和查看相关配置

（1）测试每一个对接的三层端口是否能够正常通信。

sw-h2#ping 192.168.1.13；如果通信不正常，首先要检查端口是否启动，例如：

```
sw-h2#show interfaces f1/0
FastEthernet1/0 is up, line protocol is up
……
```

如果端口没有启动，进入端口配置模式，用 no shutdown 将端口启动，如果线路协议没有启起，则要查看对端的端口配置；其次看两端端口的速率是否一致（不要用 speed auto），如果不一致，就用命令 speed 100 将端口速率设为一致；第三查看两端端口通信方式是不是一致，如果不一致，用命令 duplex 将两端端口设为一致，如果两端端口都支持全双工，最好都设置为全双工；第四要看端口 IP 地址是否在一个网段，同时要特别注意子网掩码是否一致，如果子网掩码不一致，通常会导致直接 ping 对端端口时是正常的，如果中间隔有其他网段或者通过其他路由器时，会出现通信不正常。

（2）检查二层端口的 Trunk 配置。

sw-j3#show interfaces f1/0 trunk

| Port  | Mode | Encapsulation | Status   | Native vlan |
|-------|------|---------------|----------|-------------|
| fa1/0 | on   | 802.1q        | trunking | 1           |

……

看 Trunk 协议是否正常启动，如果没有启动，首先看端口是否配置 Trunk 协议；其次查看端口是否启动；第三查看端口封装的协议是否一致。

（3）检查 VTP 协议的配置，要查看同一个域的 VTP 配置是否相同。

sw-h2# show vtp status

| | |
|---|---|
| VTP Version | : 2 |
| Configuration Revision | : 2 |
| Maximum VLANs supported locally | : 256 |
| Number of existing VLANs | : 9 |
| VTP Operating Mode | : Server |
| VTP Domain Name | : sw-h2 |
| VTP Pruning Mode | : Enabled |
| VTP V2 Mode | : Enabled |
| VTP Traps Generation | : Disabled |
| MD5 digest | : 0x0E 0x3A 0x10 0x42 0x6C 0xC3 0x46 0x29 |

Configuration last modified by 0.0.0.0 at 3-1-02 02:36:57
Local updater ID is 192.168.8.1 on interface Vl8 (lowest numbered VLAN interface found)

sw-j3#show vtp status

| | |
|---|---|
| VTP Version | : 2 |
| Configuration Revision | : 2 |
| Maximum VLANs supported locally | : 256 |
| Number of existing VLANs | : 9 |

| | |
|---|---|
| VTP Operating Mode | : Client |
| VTP Domain Name | : sw-h2 |
| VTP Pruning Mode | : Enabled |
| VTP V2 Mode | : Enabled |
| VTP Traps Generation | : Disabled |
| MD5 digest | : 0x0E 0x3A 0x10 0x42 0x6C 0xC3 0x46 0x29 |

Configuration last modified by 0.0.0.0 at 3-1-02 02:36:57

同时要注意在配置 VTP 协议时，如果配置了 VTP 密码，在同一个域中一定要保持密码一致。

（4）查看每个三层交换机和路由器上的路由配置是否正常。

```
sw-h2#show ip route
Codes: C - connected, S - static, R - RIP, M - mobile, B - BGP
       D - EIGRP, EX - EIGRP external, O - OSPF, IA - OSPF inter area
       N1 - OSPF NSSA external type 1, N2 - OSPF NSSA external type 2
       E1 - OSPF external type 1, E2 - OSPF external type 2
       i - IS-IS, su - IS-IS summary, L1 - IS-IS level-1, L2 - IS-IS level-2
       ia - IS-IS inter area, * - candidate default, U - per-user static route
       o - ODR, P - periodic downloaded static route

Gateway of last resort is 0.0.0.0 to network 0.0.0.0

C    192.168.8.0/24 is directly connected, Vlan8
C    192.168.9.0/24 is directly connected, Vlan9
C    192.168.10.0/24 is directly connected, Vlan10
C    192.168.11.0/24 is directly connected, Vlan11
     192.168.1.0/30 is subnetted, 1 subnets
C       192.168.1.12 is directly connected, FastEthernet1/0
S*   0.0.0.0/0 is directly connected, FastEthernet1/0    #默认路由
```

查看路由一定要和网络拓扑结构以及 IP 地址规划表对应起来，像上面交换机 sw-h2 的路由，以 C 开头的是直连路由，以 S* 开头的是默认路由，因为交换机 sw-h2 下面的接入层交换机 sw-j3 和 sw-j4 共有四个 vlan，sw-j3 上有 vlan 8 和 vlan 9，sw-j4 上有 vlan 10 和 vlan 11，sw-h2 上配置有四个 vlan，同时四个 vlan 都配置有 IP 地址，这四个 vlan 的地址是作为四个 vlan 下主机的网关。第五条直连路由是 sw-h2 与 sw-k 连接的网段的路由。静态默认路由是 vlan 8、vlan 9、vlan 10、vlan 11 四个 vlan 访问其他网段的数据包的提供路由。

（5）首先查看 NAT 路由器的配置，测试 NAT 路由器。

```
NAT#show ip nat Statistics
Total active translations: 0 (0 static, 0 dynamic; 0 extended)
Outside interfaces:   #外部接口
   FastEthernet1/0
Inside interfaces:    #内部接口
   FastEthernet1/1
Hits: 0  Misses: 0
CEF Translated packets: 0, CEF Punted packets: 0
Expired translations: 0
Dynamic mappings:
-- Inside Source    #内部源地址转换规则
[Id: 1] access-list 1 interface FastEthernet1/0 refcount 0
Queued Packets: 0
```

然后在 sw-k 上通过 ping firewall 防火墙的 f1/0 端口地址，同时要查看 NAT 路由器上的转换记录，例如：

```
NAT#    sho ip nat translations
Pro Inside global       Inside local         Outside local       Outside global
icmp 10.1.2.6:9         192.168.1.5:9        10.1.2.5:9          10.1.2.5:9
udp 10.1.2.6:49209      192.168.1.5:49209    10.1.2.5:33489      10.1.2.5:33489
udp 10.1.2.6:49210      192.168.1.5:49210    10.1.2.5:33490      10.1.2.5:33490
```

（6）最后进行网络整体调试。

在每一个网段 ping 其他网段的主机，看通信是否正常，如果不正常，在通过上面的方法进行相应的调试，直到整个网络中每两个网段之间的通信都正常为止。

# 实验十四　路由器 SSH 配置

通过 SSH 登录远程路由器，提高远程登录路由器的安全性，从而提高网络的安全性。

## 一、实验目的

（1）了解 SSH 的工作原理；
（2）熟练掌握路由器远程登录 SSH 的配置步骤。

## 二、实验内容

（1）了解 SSH 的工作原理；
（2）熟练掌握路由器远程登录 SSH 的配置步骤。

## 三、实验原理

安全外壳协议（Secure Shell Protocol，SSH）是一种在不安全网络上提供安全远程登录及其他安全网络服务的协议。Secure Shell 又可记为 SSH，最初是 UNIX 系统上的一个程序，后来又迅速扩展到其他操作平台。SSH 是一个很好的应用程序，在正确使用时，它可以弥补网络中的漏洞。除此以外，SSH 之所以被广泛地应用，还有以下原因：SSH 客户端适用于多种平台，几乎所有的 UNIX 平台（包括 HP-UX、Linux、AIX、Solaris、Digital UNIX、Irix、SCO，以及其他平台）。而且，已经有一些客户端可以运行于 UNIX 操作平台以外，包括 OS/2、VMS、BeOS、Java、Windows 95/98 和 Windows NT。这样就可以在几乎所有的平台上运行 SSH 客户端程序了。对非商业用途，它是免费的。许多 SSH 版本可以获得源代码，并且只要不用于商业目的，都可以免费得到。而且，UNIX 版本也提供了源代码，这就意味着任何人都可以对它进行修改。但是，如果选择它用于商业目的，那么无论使用何种版本的 SSH，都得确认已经注册并获得了相应权限。绝大多数 SSH 的客户端和守护进程都有一些注册限制。唯一的 SSH 通用公共注册（General Public License，GPL）版本是 LSH，它目前还是测试版。通过 Internet 传送密码安全可靠，这是 SSH 被认可的优点之一。如果考察一下接入 ISP（Internet Service Provider，Internet 服务供应商）或大学的方法，一般都是采用 Telnet 或 POP 邮件客户进程。因此，每当要进入自己的账号时，你输入的密码将会以明码方式发送（即没有保护，直接可读），这就给攻击者一个盗用你账号的机会。

但是因为受版权和加密算法的限制，现在很多人都转而使用 OpenSSH。OpenSSH 是 SSH 的替代软件，而且是免费的，可以预计将来会有越来越多的人使用它来代替 SSH。SSH 是由客户端和服务端的软件组成的，有两个不兼容的版本：1.x 和 2.x。用 SSH 2.x 的客户程序是不能连接到 SSH 1.x 的服务程序上去的。OpenSSH 2.x 同时支持 SSH 1.x 和 SSH 2.x。

SSH 主要由三部分组成：

（1）传输层协议（SSH-TRANS）提供了服务器认证、保密性及完整性。此外它有时还提供压缩功能。SSH-TRANS 通常运行在 TCP/IP 连接上，也可能用于其他可靠数据流上。SSH-TRANS 提供了强大的加密技术、密码主机认证及完整性保护。该协议中的认证基于主机，并且该协议不执行用户认证。更高层的用户认证协议可以设计为在此协议之上。

（2）用户认证协议（SSH-USERAUTH）用于向服务器提供客户端用户鉴别功能。它运行在传输层协议 SSH-TRANS 上面。当 SSH-USERAUTH 开始后，它从低层协议那里接收会话标识符（从第一次密钥交换中获得的哈希值）。会话标识符唯一标识此会话，并且适用于证明私钥的所有权。SSH-USERAUTH 也需要知道低层协议是否提供保密性保护。

（3）连接协议（SSH-CONNECT）将多个加密隧道分成逻辑通道。它运行在用户认证协议上。它提供了交互式登录话路、远程命令执行、转发 TCP/IP 连接和转发 X11 连接。

一旦建立一个安全传输层连接，客户机就发送一个服务请求。当用户认证完成之后，会发送第二个服务请求。这样就允许新定义的协议可以与上述协议共存。连接协议提供了用途广泛的各种通道，有标准的方法用于建立安全交互式会话外壳及转发（"隧道技术"）专有 TCP/IP 端口和 X11 连接。

通过使用 SSH，可以把所有传输的数据进行加密，这样"中间人"这种攻击方式就不可能实现了，而且也能够防止 DNS 欺骗和 IP 欺骗。使用 SSH，还有一个额外的好处就是传输的数据是经过压缩的，所以可以加快传输的速度。SSH 有很多功能，它既可以代替 Telnet，又可以为 FTP、PoP 甚至为 PPP 提供一个安全的"通道"。

SSH 分为两部分：服务端部分和客户端部分。

服务端是一个守护进程（demon），它在后台运行并响应来自客户端的连接请求。服务端一般是 SSHD 进程，提供了对远程连接的处理，一般包括公共密钥认证、密钥交换、对称密钥加密和非安全连接。

客户端包含 SSH 程序以及像 SCP（远程拷贝）、SLOGIN（远程登录）、SFTP（安全文件传输）等其他的应用程序。

它们的工作机制大致是本地的客户端发送一个连接请求到远程的服务端，服务端检查申请的包和 IP 地址，再发送密钥给 SSH 的客户端，本地再将密钥发回给服务端，自此连接建立。

从客户端来看，SSH 提供以下两种级别的安全验证。

第一种级别（基于口令的安全验证）：只要知道自己账号和口令，就可以登录到远程主机。所有传输的数据都会被加密，但是不能保证正在连接的服务器就是你想连接的服务器。可能会有其他服务器在冒充真正的服务器，也就是受到"中间人"这种方式的攻击。

第二种级别（基于密钥的安全验证）：需要依靠密钥，也就是必须为自己创建一对密钥，并把公用密钥放在需要访问的服务器上。如果要连接到 SSH 服务器上，客户端软件就会向服务器发出请求，请求用密钥进行安全验证。服务器收到请求之后，先在该服务器的目录下寻找公用密钥，然后把它和发送过来的公用密钥进行比较。如果两个密钥一致，服务器就用公用密钥加密"质询"（challenge），并把它发送给客户端软件。客户端软件收到"质询"之后，就可以用私人密钥解密，再把它发送给服务器。用这种方式时，你必须知道自己密钥的口令。但是，与第一种级别相比，第二种级别不需要在网络上传送口令。

第二种级别不仅加密所有传送的数据，而且"中间人"这种攻击方式也是不可能的。但是整个登录的过程可能需要 10 秒。

## 四、实验需要掌握的命令

Router(config)#ip domain-name cisco.com
#为路由器配置一个域名

Router (config)#crypto key generate rsa general-keys modulus *number*
#生成一个 RSA 算法的密钥

Router (config)#ip ssh time-out 120
#设置 SSH Server 的用户认证超时时间

Router (config)#ip ssh authentication 4
#设置 SSH Server 认证重复次数

Router (config-line)#transport input ssh
#设置 VTY 的登录模式为 SSH，默认情况下是 all（即允许所有登录）

Router (config)#aaa new-model
#启用 AAA

Router (config)#aaa authentication login default local
#启用 AAA 认证，设置在本地服务器上进行认证

Router (config)#username *username* password *password*
#创建一个用户 *username*，并设置其密码为 *password*，用于 SSH 客户端登录

Router (config-line)#login authentication default
#设置使用 AAA 的 default 来进行认证

Router #show crypto key mypubkey rsa
#查看加密的公开密钥和私有密钥

## 五、实验拓扑图和拓扑文件

（1）网络拓扑图。

网络拓扑图如图 14-1 所示。

图 14-1　网络拓扑图

（2）网络文件。

```
#ssh.net 文件开始
[localhost]
    [[3745]]
    image = ..\c3745-adventerprisek9-mz.124-16.bin
    ram = 160
    [[ROUTER R1]]
    idlepc = 0x6064179c
    model = 3745
    F0/0 = NIO_gen_eth:\Device\NPF_{E2B31736-C716-4215-A9CB-ADA78C760F52}
#ssh.net 文件结束
```

## 六、实验配置步骤

router>
router>enable

```
router#configure terminal
router(config)#enable password 123456
#如果不设置 enable 密码,则远程登录的用户无法进入特权模式
router(config)#hostname ssh
ssh(config)#ip domain-name cisco.com
ssh(config)#crypto key generate rsa general-keys modulus 1024
ssh(config)#aaa new-model
ssh(config)#aaa authentication login default local
ssh(config)#username cisco password cisco
ssh(config)#ip ssh time-out 120
ssh(config)#ip ssh authentication-retries 4
ssh(config)#interface FastEthernet0/0
ssh(config-if)#ip address 192.168.7.157 255.255.255.0
ssh(config-if)#duplex full
ssh(config-if)#speed 100
ssh(config-if)#no shutdown
ssh(config)#line vty 0 4
ssh(config-line)#transport input ssh
ssh(config-line)#end
ssh#write
```

## 七、实验测试和查看相关配置

(1) 查看加密的公开密钥和私有密钥。

```
ssh#show crypto key mypubkey rsa
% Key pair was generated at: 00:02:14 UTC Mar 1 2002
Key name: R1.cisco.com
 Usage: General Purpose Key
 Key is not exportable.
 Key Data:
  30819F30 0D06092A 864886F7 0D010101 05000381 8D003081 89028181 009805ED
  A288857F 453A8524 CB9252A8 A17D1C4F FC36D599 1C3DE62A 900ED528 A5EA9526
  BE048D84 02AC3C2B A7D0DF32 8993C0F2 D0DE1F47 9DF35A8C 57229098 1AFACD40
  D644F291 16F8AF65 9623E39A B9D983E6 54ED4142 BAA65E74 3E648F39 1554660C
  DE123D99 001D0EBB E901E631 EEF5ECD9 FE88FC83 A52D7FA8 FA562BD9 A7020301 0001
% Key pair was generated at: 01:02:16 UTC Mar 1 2002
Key name: R1.cisco.com.server
 Usage: Encryption Key
 Key is not exportable.
 Key Data:
  307C300D 06092A86 4886F70D 01010105 00036B00 30680261 00EAC67B 648D0D82
  6D966826 91239D1D 0A0D9A6A 8ED22351 980AE97D BC90AFA5 1A7FA3B1 F047AB60
  A3CE5598 A269A6B2 636E117C A9BA4820 E7E521C9 21EF79A7 ED7F65CD 2288EF89
  76703AD3 BA70F657 10D6C4F7 CABF576D E2D8F36C FB23A733 6B020301 0001
```

(2) 测试 SSH 登录。

运行 SSH 的客户端程序 putty.exe,输入路由器的 IP 地址,如图 14-2 所示。

然后单击 Open 按钮,进入如图 14-3 所示的窗口,输入用户名和密码进入用户模式,最后输入 enable 命令和密码,就进入到特权模式。

(3) 查看数据包是否加密。

在 dynagen.exe 程序运行窗口的命令行输入 capture router f0/0 ssh.cap;然后通过 putty.exe

客户端登录路由器；最后用协议分析器 wireshark.exe 程序查看数据包是否加密。结果如图 14-4 所示（其中灰色的部分为加密的数据）。

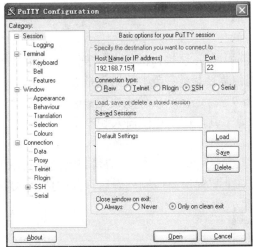

图 14-2　SSH 客户端登录程序

图 14-3　登录路由器

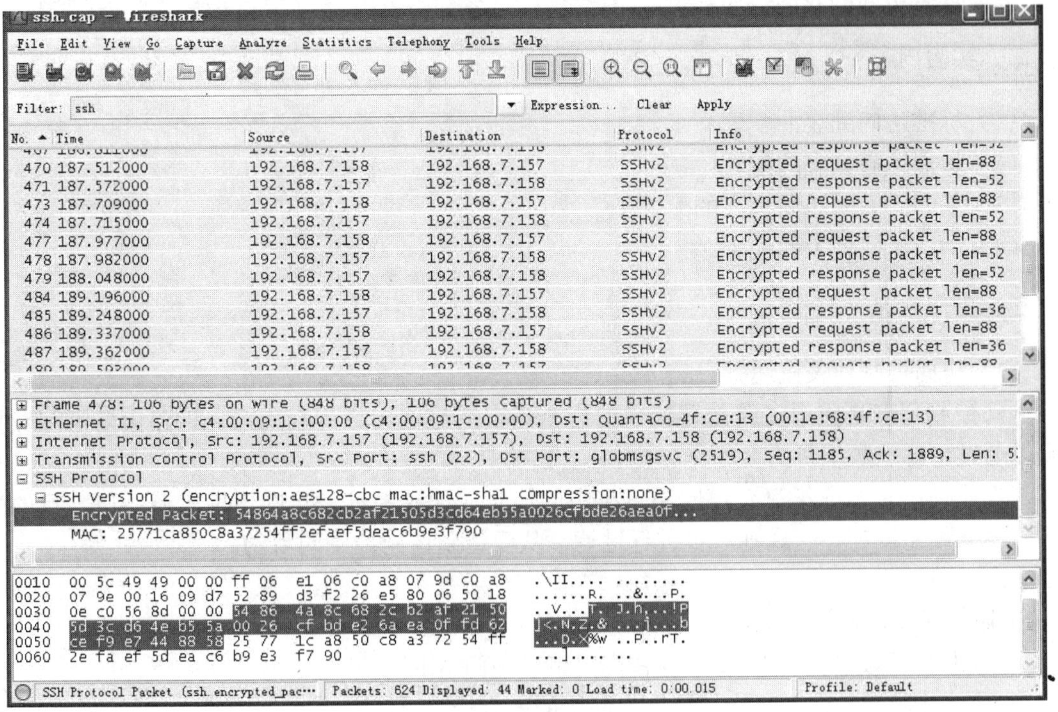

图 14-4　查看抓取的数据包

## 八、注意问题

（1）要设置 enable 密码，如果不设置 enable 密码，则远程登录的用户无法进入特权模式。
（2）路由器的镜像文件应该是高级版本，否则路由器不支持隧道加密。

# 实验十五  访问列表

访问控制列表（Access Control List，ACL）是应用在三层接口的指令列表，这些指令列表用来告诉网络设备哪些数据包可以接收、哪些数据包需要拒绝。至于数据包是被接收还是被拒绝，可以由类似于源地址、目的地址、端口号、协议等特定指示条件来决定。通过灵活地增加访问控制列表，ACL 可以当作一种网络控制的有力工具，用来过滤流入和流出路由器接口的数据包。

建立访问控制列表后，可以限制网络流量，提高网络性能，对通信流量起到控制的作用，这也是对网络访问的基本安全手段。在路由器的接口上配置访问控制列表后，可以对入站接口、出站接口及通过路由器中继的数据包进行安全检测。

## 一、实验目的

（1）掌握访问列表实验安全性的配置；
（2）理解访问列表的作用。

## 二、实验内容

（1）标准访问列表的配置；
（2）扩展访问列表的配置；
（3）基于时间的访问列表的配置；
（4）动态访问列表的配置。

## 三、工作原理

（1）访问控制列表概述。

访问控制列表是应用在路由器接口的指令列表。这些指令列表用来告诉路由器哪些数据包可以通过，哪些数据包需要拒绝。

工作原理：它读取第三及第四层包头中的信息，如源地址、目的地址、源端口、目的端口等。根据预先设定好的规则对包进行过滤，从而达到访问控制的目的。

实际应用：

1）阻止某个网段访问服务器。
2）阻止 A 网段访问 B 网段，但 B 网段可以访问 A 网段。
3）禁止某些端口进入网络，可提高网络的安全性。

（2）标准 ACL。

标准访问控制列表只检查被路由器路由的数据包的源地址。若使用标准访问控制列表禁用某网段，则该网段下所有主机以及所有协议都被禁止。如禁止了 A 网段，则 A 网段下所有的主机都不能访问服务器，而 B 网段下的主机却可以。

用 1~99 和 1300~1999 之间数字作为表号。

标准 ACL 一般用于局域网，所以最好将其应用在离目的地址最近的地方。

标准访问控制列表的工作原理如图 15-1 所示（每当数据包进入路由器的每个接口，都会进行以下流程）。

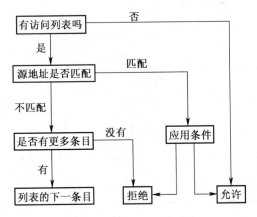

图 15-1　标准访问列表流程

**注意**：当配置访问控制列表时，顺序很重要。要确保按照从具体到普遍的次序来排列条目。

（3）扩展 ACL。

扩展访问控制列表对数据包源地址、目的地址、源端口、目的端口都进行检查。若使用扩展访问控制列表禁止某网段而访问其他网段，则 A 网段下所有主机不能访问 B 网段，而 B 网段下的主机可以访问 A 网段。

用 100～199 和 2000～2699 之间数字作为表号。

扩展 ACL 一般用于外网，所以最好将其应用在离源地址最近的地方。

扩展访问控制列表的工作原理如图 15-2 所示（每当数据包进入路由器的每个接口，都会进行以下流程）：

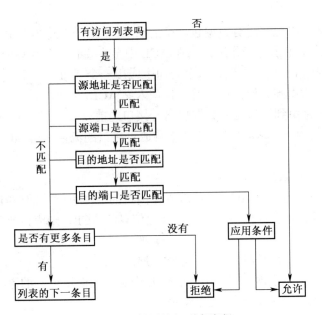

图 15-2　扩展访问列表流程

常用端口及所属协议如表 5-1 所示。

表 15-1 常用端口及所属协议

| 端口号 | 关键字 | 描述 | TCP/UDP |
| --- | --- | --- | --- |
| 0-4 | | 未分配 | |
| 20 | ftp（data） | 文件数据传输协议 | TCP |
| 21 | ftp | 文件传输协议 | TCP |
| 23 | telnet | 终端连接 | TCP |
| 25 | smtp | 简单邮件传输协议 | TCP |
| 42 | nameserver | 主机名服务器 | UDP |
| 53 | domain | 域名服务器（DNS） | TCP/UPD |
| 80 | www | WWW 服务 | TCP |
| 8089 | | 远程控制 | TCP |

（4）基本规则、准则和限制。

1）ACL 语句按名称或编号分组；

2）每条 ACL 语句都只有一组条件和操作，如果需要多个条件或多个行动，则必须生成多个 ACL 语句；

3）如果一条语句的条件中没有找到匹配，则处理列表中的下一条语句；

4）如果在 ACL 组的一条语句中找到匹配，则不再处理后面的语句；

5）如果处理了列表中的所有语句而没有指定匹配，不可见到的隐式拒绝语句拒绝该数据包；

6）由于在 ACL 语句组的最后隐式拒绝，所以至少要有一个允许操作，否则，所有数据包都会被拒绝；

7）语句的顺序很重要，约束性最强的语句应该放在列表的顶部，约束性最弱的语句应该放在列表的底部；

8）一个空的 ACL 组允许所有数据包，空的 ACL 组已经在路由器上被激活，但不包含语句的 ACL，要使隐式拒绝语句起作用，则在 ACL 中至少要有一条允许或拒绝语句；

9）只能在每个接口、每个协议、每个方向上应用一个 ACL；

10）在数据包被路由到其他接口之前，处理入站 ACL；

11）在数据包被路由到接口且在数据包离开接口之前，处理出站 ACL；

12）当 ACL 应用到一个接口时，这会影响通过接口的流量，但 ACL 不会过滤路由器本身产生的流量。

（5）ACL 放置在什么位置。

1）只过滤数据包源地址的 ACL 应该放置在离目的地尽可能近的地方；

2）过滤数据包的源地址、目的地址以及其他信息的 ACL，则应该放在离源地址尽可能近的地方；

3）只过滤数据包中的源地址的 ACL 有两个局限性：

①即使 ACL 应用到路由器 C 的 E0，从 ACL 中定义的源地址来的流量都将被禁止访问路由器 A 的 E0 网段的任何资源，包括数据库服务器。

②流量要经过所有到达目的地的途径，它在即将到达目的地时被丢弃，这是对带宽的浪费。

**四、实验需要掌握的命令**

（1）标准 ACL 的配置。

router(config)#access-list *access-list-number* deny *subnet wildcard-bits*

#禁止某个网段或某个 IP，其中 *access-list-number* 为访问列表的编号，范围是 1~99 和 1300~1999；*subnet* 为网段地址或主机地址；*wildcard-bits* 为子网掩码的反码

router(config)#access-list *access-list-number* permit *subnet wildcard-bits*

#允许某个网段或某个 IP，其中 *access-list-number* 为访问列表的编号，范围是 1~99 和 1300~1999；*subnet* 为网段地址或主机地址；*wildcard-bits* 为子网掩码的反码

router(config)#access-list *access-list-number* permit any

#允许所有 IP 的数据包通过

router（config-if）#ip access-group *access-list-number* in/out

#在端口模式下应用编号为 *access-list-number* 的访问列表，in/out 表示数据包通过端口的方向

（2）扩展访问控制列表的配置。

router(config)#access-list *access-list-number* [dynamic *name* [timeout *time*]] deny *protocol source-subnet wildcard-bits* [*match port-number*] *destination-subnet wildcard-bits* [*match port-number*] [time-range *time-range-name*]

#源 IP 地址在 *source-subnet wildcard-bits* 网段，目的 IP 地址在 *destination-subnet wildcard-bits* 网段，并且端口号和协议匹配的数据包被禁止。其中 *access-list-number* 为访问列表的编号，范围是 100~199 和 2000~2699；dynamic 为可选项，配置动态访问列表关键字，*name* 为名称，timeout *time* 是制定超时的时间值，单位为秒；*protocol* 为协议，有<0-255>（协议号）、ahp、eigrp、esp、gre、icmp、igmp、ip、ipinip、nos、ospf、pcp、pim、tcp、udp；*wildcard-bits* 为子网掩码的反码；*match* 为匹配符，有 ack、dscp、eq、establishe、fin、fragments、gt、log、log-input、lt、neq、precedence、psh、range、rst、syn、time-range、tos、urg；time-range 为时间范围命令，*time-range-name* 为时间范围的名称

router（config）# access-list *access-list-number* [dynamic *name* [timeout *time*]] permit *protocol source-subnet wildcard-bits* [*match port-number*] *destination-subnet wildcard-bits* [*match port-number*] [time-range *time-range-name*]

#源 IP 地址在 *source-subnet wildcard-bits* 网段，目的 IP 地址在 *destination-subnet wildcard-bits* 网段，并且端口号和协议匹配的数据包被允许

router（config-if）#ip access-group *access-list-number* in/out

#在端口模式下应用编号为 *access-list-number* 的访问列表，in/out 表示数据包通过端口的方向

router#show access-lists

#查看访问控制列表

router(config)#time-range *time-range-name*

#进入时间范围配置模式，*time-range-name* 为时间范围的名称

router(config-time-range)# periodic *day start-time* to *end-time*

#配置相对时间列表命令，*day* 为 Friday、Monday、Saturday、Sunday、Thursday、Tuesday、

Wednesday、daily、weekdays、weekend 中的一个或多个的组合；start-time 为开始时间，end-time 为结束时间

router(config-time-range)# absolute [start *start-time*] [to] [end *end-time*]

#配置绝对时间列表命令，*start-time* 为起始时间，*end-time* 为结束时间，其中的起始时间和结束时间只有一个。如 absolute start *start-time* 或 absolute end *end-time* 都是可以的

router (config-line)autocommand access-enable host timeout *time*

#在通过虚拟终端登录时，路由器自动执行命令 access-enable host timeout *time*，使动态访问列表生效，其中 *time* 为时间，单位是分钟

### 五、实验拓扑图和网络文件

（1）网络拓扑图。

网络拓扑图如图 15-3 所示。

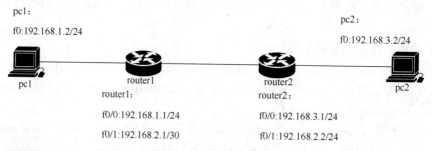

图 15-3　网络拓扑图

（2）网络文件。

```
# access-list.net 文件开始
autostart = false
ghostios = true
sparsemem = true

[localhost]

    [[3660]]
    image = ..\c3660-is.bin
    ram = 160

    [[1700]]
    image = ..\C1700-Y.BIN
    ram = 160

    [[ROUTER Router1]]
    idlepc = 0x60719b98
    model = 3660
    f0/0 = pc1 f0
    f0/1 = Router2 f0/1


    [[router Router2]]
```

```
            idlepc = 0x60719b98
            model = 3660
            f0/0 = pc2 f0

            [[router pc1]]
            model = 1720
            idlepc = 0x8014e4ec

            [[router pc2]]
            model = 1720
            idlepc = 0x8014e4ec
# access-list.net 文件结束
```

## 六、实验配置步骤

（1）基本配置。

1）路由器 router1 的基本配置。

```
router>enable
router#configure terminal
router(config)#hostname router1
router1(config)#interface FastEthernet0/0
router1(config-if)#ip address 192.168.1.1 255.255.255.0
router1(config-if)#duplex full
router1(config-if)#speed 100
router1(config-if)#no shutdown
router1(config-if)#interface FastEthernet0/1
router1(config-if)#ip address 192.168.2.1 255.255.255.252
router1(config-if)#duplex full
router1(config-if)#speed 100
router1(config-if)#no shutdown
router1(config-if)#exit
router1(config)#ip route 0.0.0.0 0.0.0.0 FastEthernet0/1
router1(config)#end
router1#write
```

2）路由器 router2 的基本配置。

```
router>enable
router#configure terminal
router(config)#hostname router2
router2(config)#interface FastEthernet0/0
router2(config-if)#ip address 192.168.3.1 255.255.255.0
router2(config-if)#duplex full
router2(config-if)#speed 100
router2(config-if)#no shutdown
router2(config-if)#interface FastEthernet0/1
router2(config-if)#ip address 192.168.2.2 255.255.255.252
router2(config-if)#duplex full
router2(config-if)#speed 100
router2(config-if)#no shutdown
```

```
router2(config-if)#exit
router2(config)#ip route 0.0.0.0 0.0.0.0 FastEthernet0/1
router2(config)#end
router2#write
```

3）主机 pc1 的基本配置。

```
router>enable
router#configure terminal
router(config)#hostname pc1
pc1(config)#interface FastEthernet0
pc1(config-if)#ip address 192.168.1.2 255.255.255.0
pc1(config-if)# duplex full
pc1(config-if)#no shutdown
pc1(config-if)#exit
pc1(config)#no ip routing
pc1(config)#ip default-gateway 192.168.1.1
pc1 (config)#end
pc1 #write
```

4）主机 pc2 的基本配置。

```
router>enable
router#configure terminal
router(config)#hostname pc2
pc2(config)#interface FastEthernet0
pc2(config-if)#ip address 192.168.3.2 255.255.255.0
pc2(config-if)# duplex full
pc2(config-if)#no shutdown
pc2(config-if)#exit
pc2(config)#no ip routing
pc2(config)#ip default-gateway 192.168.3.1
pc2(config)#end
pc2#write
```

（2）标准访问列表配置。

禁止主机 pc1 ping 主机 pc2

```
router2 (config)#access-list 1 deny 192.168.1.2
router2 (config)#access-list 1 permit any
router2 (config)# interface FastEthernet0/0
router2 (config-if)# ip access-group 1 out
```

（3）扩展访问列表配置。

禁止主机 pc1 ping 主机 pc2，同时允许主机 pc2 ping 主机 pc1

```
router2 (config)#access-list 100 deny    icmp host 192.168.1.2 host 192.168.3.2 echo
router2 (config)#access-list 100 permit ip any any
router2 (config)# interface FastEthernet0/0
router2 (config-if)#ip access-group 100 out
```

此配置也可以在 router1 路由器上配置，配置如下：

```
router1 (config)#access-list 100 deny icmp host 192.168.1.2 host 192.168.3.2 echo
router1 (config)#access-list 100 permit ip any any
router1 (config)# interface FastEthernet0/0
router1 (config-if)#ip access-group 100 in
```

（4）基于时间的访问列表配置。

在星期一到星期五凌晨 0:00 到 8:00、星期一到星期四晚上 22:00 到 23:59 和星期天晚上 22:00 到 23:59，不允许任何的数据包进入路由器 router2 的 f0/0 端口，配置如下：

router2(config)#time-range student
router2(config-time-range)#periodic weekdays 0:00 to 8:00
router2(config-time-range)#periodic Monday Tuesday Wednesday Thursday Sunday 22:00 to 23:59
router2(config-time-range)#exit
router2(config)#access-list 102 deny ip any any time-range student
router2(config)#access-list 102 permit ip any any
router2 (config)# interface FastEthernet0/0
router2 (config-if)#ip access-group 102 in

（5）动态访问列表配置。

pc2 没有登录 router2 时，pc1 无法 ping 主机 pc2，访问列表中动态访问列表没有生效；在主机 pc2 登录 router2 进行一次认证之后，pc1 可以 ping 主机 pc2，访问列表中动态访问列表生效。

router2 (config)# access-list 101 permit tcp host 192.168.3.2 any eq telnet
router2 (config)# access-list 101 dynamic abc timeout 60 permit ip any any
router2 (config)# interface FastEthernet0/0
router2 (config-if)# ip access-group 101 in
router2 (config-if)#exit
router2 (config)# line vty 0 4
router2 (config-line)# password 0 123456
router2 (config-line)# login
router2 (config-line)#autocommand access-enable host timeout 1

### 七、实验测试和查看相关配置

（1）标准访问列表配置测试。

在 pc1 上无法 ping 主机 pc2，但是可以 ping 其他所有的端口和主机，结果如下：

pc1> ping 192.168.3.2
Type escape sequence to abort.
Sending 5, 100-byte ICMP Echos to 192.168.3.2, timeout is 2 seconds:
U.U.U
Success rate is 0 percent (0/5)
pc1> ping 192.168.3.1
Type escape sequence to abort.
Sending 5, 100-byte ICMP Echos to 192.168.3.1, timeout is 2 seconds:
!!!!!
Success rate is 100 percent (5/5), round-trip min/avg/max = 36/58/80 ms

在 pc2 上也无法 ping 主机 pc1
pc2#ping 192.168.1.2
Type escape sequence to abort.
Sending 5, 100-byte ICMP Echos to 192.168.1.2, timeout is 2 seconds:
.....
Success rate is 0 percent (0/5)

（2）扩展访问列表配置测试。

在 pc1 上无法 ping 主机 pc2

pc1>ping 192.168.3.2

Type escape sequence to abort.

Sending 5, 100-byte ICMP Echos to 192.168.3.2, timeout is 2 seconds:

U.U.U

Success rate is 0 percent (0/5)

在 pc2 上可以 ping 主机 pc1

pc2#ping 192.168.1.2

Type escape sequence to abort.

Sending 5, 100-byte ICMP Echos to 192.168.1.2, timeout is 2 seconds:

!!!!!

Success rate is 100 percent (5/5), round-trip min/avg/max = 36/54/96 ms

（3）基于时间的访问列表配置测试。

在路由器 router2 通过 router2#clock set 7:00:00 5 oct 2010 命令将时间修改为 2010 年 10 月 5 日的 7:00，则在 pc2 上无法 ping 任何主机。结果如下：

pc2>ping 192.168.1.1

Type escape sequence to abort.

Sending 5, 100-byte ICMP Echos to 192.168.1.1, timeout is 2 seconds:

U.U.U

Success rate is 0 percent (0/5)

在路由器 router2 通过 router2#clock set 9:00:00 5 oct 2010 命令将时间修改为 2010 年 10 月 5 日的 9:00，则在 pc2 上可以 ping 任何主机。结果如下：

pc2>ping 192.168.1.1

Type escape sequence to abort.

Sending 5, 100-byte ICMP Echos to 192.168.1.1, timeout is 2 seconds:

!!!!!

Success rate is 100 percent (5/5), round-trip min/avg/max = 40/55/88 ms

（4）动态访问列表配置测试。

pc2 没有登录 router2 时，pc1 无法 ping 主机 pc2，访问列表中动态访问列表没有生效；在主机 pc2 登录 router2 进行一次认证之后，pc1 可以 ping 主机 pc2，访问列表中动态访问列表生效。结果如下：

**pc2 登录 router2 进行认证之前：**

pc1#ping 192.168.3.2

Type escape sequence to abort.

Sending 5, 100-byte ICMP Echos to 192.168.3.2, timeout is 2 seconds:

.....

Success rate is 0 percent (0/5)

router2#show access-lists

Extended IP access list 101

    10 permit tcp host 192.168.3.2 any eq telnet (435 matches)

    20 Dynamic abc permit ip any any

**pc2 通过 telnet 登录 router2 进行认证之后：**

pc1#ping 192.168.3.2

Type escape sequence to abort.

Sending 5, 100-byte ICMP Echos to 192.168.3.2, timeout is 2 seconds:

!!!!!
Success rate is 100 percent (5/5), round-trip min/avg/max = 36/50/64 ms
router2#show access-lists
Extended IP access list 101
    10 permit tcp host 192.168.3.2 any eq telnet (486 matches)
    20 Dynamic abc permit ip any any
      permit ip host 192.168.3.2 any (5 matches) (time left 47)

## 八、注意问题

（1）区分标准访问列表和扩展访问列表的不同。

（2）将访问列表应用到端口上时，要注意方向。

（3）动态访问列表配置中，要理解 autocommand access-enable host timeout 1 命令的含义，它是通过 telnet 登录到路由器时，路由器会在特权模式下自动执行 access-enable host timeout 1 命令，同时注意，此命令中 timeout 后面的时间单位是分钟。

（4）当配置访问控制列表时，顺序很重要。要确保按照从具体到普遍的次序来排列条目。

（5）删除访问控制列表就会将同一个序号的访问控制列表全部删除，不能只删除其中的一条。

# 实验十六  RIP 路由协议的配置

路由选择协议主要是运行在路由器上的协议，主要用来进行路径选择。RIP（Routing Information Protocol，路由信息协议）是应用较早、使用较普遍的内部网关协议（Interior Gateway Protocol，IGP），适用于小型同类网络的一个自治系统（AS）内的路由信息的传递。RIP 协议是基于距离矢量算法（Distance Vector Algorithm，DVA）的。它使用"跳数"（metric）来衡量到达目标地址的路由距离，是一个用于路由器和主机间交换路由信息的距离向量协议，目前最新的版本为 v4，即 RIPv4。

## 一、实验目的

（1）理解 RIP 路由协议的工作原理；
（2）掌握 RIP 路由协议的配置。

## 二、实验内容

（1）RIP 的基本配置；
（2）RIP 的定时器的配置；
（3）RIP 的路由汇总的配置；
（4）RIP 的组播更新的配置；
（5）RIP 路由的交换安全配置。

## 三、实验原理

（1）RIP 工作原理。

RIP 协议是基于 Bellham-Ford（距离向量）算法，此算法于 1969 年被用于计算机路由选择，正式协议首先是由 Xerox 于 1970 年开发的，当时是作为 Xerox 的 Networking Services（NXS）协议簇的一部分。由于 RIP 实现简单，迅速成为使用范围最广泛的路由协议。

路由器的关键作用是用于网络的互连，每个路由器与两个以上的实际网络相连，负责在这些网络之间转发数据报。在讨论 IP 进行选路和对报文进行转发时，我们总是假设路由器包含了正确的路由，而且路由器可以利用 ICMP 重定向机制来要求与之相连的主机更改路由。但在实际情况下，IP 进行选路之前必须先通过某种方法获取正确的路由表。在小型的、变化缓慢的互连网络中，管理者可以用手工方式来建立和更改路由表。而在大型的、迅速变化的环境下，人工更新的办法慢得不能接受。这就需要自动更新路由表的方法，即所谓的动态路由协议，RIP 协议是其中最简单的一种。

在路由实现时，RIP 作为一个系统常驻进程（daemon）存在于路由器中，负责从网络系统的其他路由器接收路由信息，从而对本地 IP 层路由表作动态的维护，保证 IP 层发送报文时选择正确的路由。同时负责广播本路由器的路由信息，通知相邻路由器做相应的修改。RIP 协议处于 UDP 协议的上层，RIP 所接收的路由信息都封装在 UDP 协议的数据报中，RIP 在 520 号 UDP 端口上接收来自远程路由器的路由修改信息，并对本地的路由表做相应的修改，同时通知其他路由器。通过这种方式达到全局路由的有效。

RIP 路由协议用"更新（UPDATES）"和"请求（REQUESTS）"这两种分组来传输信息。每个具有 RIP 协议功能的路由器每隔 30 秒用 UDP520 端口给与之直接相连的机器广播更新信息。更新信息反映了该路由器所有的路由选择信息数据库。路由选择信息数据库的每个条目由"局域网上能达到的 IP 地址"和"与该网络的距离"两部分组成。请求信息用于寻找网络上能发出 RIP 报文的其他设备。

RIP 用"路程段数"（即"跳数"）作为网络距离的尺度。每个路由器在给相邻路由器发出路由信息时，都会给每个路径加上内部距离。在图 16-1 中，路由器 3 直接和网络 C 相连。当它向路由器 2 通告网络 142.10.0.0 的路径时，它把跳数增加 1。与之相似，路由器 2 把跳数增加到 2，且通告路径给路由器 1，则路由器 2 和路由器 1 与路由器 3 所在网络 142.10.0.0 的距离分别是 1 跳和 2 跳。

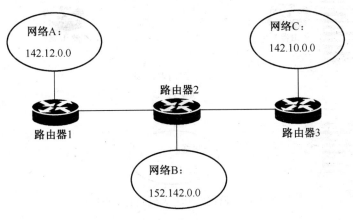

图 16-1　RIP 工作原理

然而实际的网络路由选择并不总是由跳数决定的，还要结合实际的路径连接性能综合考虑。在如图 16-2 所示的网络中，从路由器 1 到网络 C，RIP 协议将更倾向于跳数为 2 的路由器 1→路由器 2→路由器 3 的 1.5Mb/s 链路，而不是选择跳数为 1 的 56kb/s，直接的路由器 1→路由器 3 路径，因为跳数为 1 的 56kb/s 串行链路比跳数为 2 的 1.5Mb/s 串行链路慢得多。

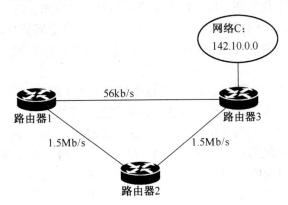

图 16-2　路由选择不仅限于"跳数"考虑

（2）路由器的收敛机制。

任何距离向量路由选择协议（如 RIP）都有一个问题，路由器不知道网络的全局情况，路由器必须依靠相邻路由器来获取网络的可达信息。由于路由选择更新信息在网络上传播慢，距离向量路由选择算法有一个慢收敛问题，这个问题将导致不一致性产生。RIP 协议使用以下机制减少因网络上的不一致带来的路由选择环路的可能性。

1）记数到无穷大机制。

RIP 协议允许最大跳数为 15，大于 15 的目的地被认为是不可达。这个数字在限制了网络大小的同时也防止了一个叫做"记数到无穷大"的问题。

记数到无穷大机制的工作原理如下（如图 16-3 所示）。

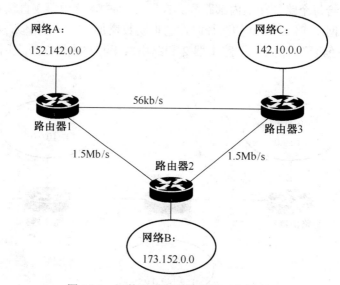

图 16-3　记数到无穷大机制的工作原理

- 现假设路由器 1 断开了与网络 A 相连，则路由器 1 丢失了与网络 A 相连的以太网接口后，产生一个触发更新送往路由器 2 和路由器 3。这个更新信息同时告诉路由器 2 和路由器 3，路由器 1 不再有到达网络 A 的路径。假设这个更新信息传输到路由器 2 被推迟了（CPU 忙、链路拥塞等），但到达了路由器 3，所以路由器 3 会立即从路由表中去掉到网络 A 的路径。

- 路由器 2 由于未收到路由器 1 的触发更新信息，并发出它的常规路由选择更新信息，通告网络 A 以 2 跳的距离可达。路由器 3 收到这个更新信息，认为出现了一条通过路由器 2 的到达网络 A 的新路径。于是路由器 3 告诉路由器 1，它能以 3 跳的距离到达网络 A。

- 在收到路由器 3 的更新信息后，就把这个信息加上 1 跳后向路由器 2 和路由器 3 同时发出，告诉它们路由器 1 可以以 3 跳的距离到达网络 A。

- 路由器 2 在收到路由器 1 的消息后，通过比较发现与原来到达网络 A 的路径不符，更新成可以以 4 跳的距离到达网络 A。这个消息会再次发往路由器 3，以此循环，直到跳数达到超过 RIP 协议允许的最大值（在 RIP 中定义为 16）。一旦一个路由器达到这个值，它将声明这条路径不可用，并从路由表中删除此路径。

由于记数到无穷大问题，路由选择信息将从一个路由器传到另一个路由器，每次段数加 1。路由选择环路问题将无限制地进行下去，除非达到某个限制，这个限制就是 RIP 的最大跳数。当路径的跳数超过 15，这条路径才从路由表中删除。

2）水平分割法。

水平分割规则如下：路由器不向路径到来的方向回传此路径。当打开路由器接口后，路由器记录路径是从哪个接口来的，并且不向此接口回传此路径。

Cisco 可以对每个接口关闭水平分割功能。这个特点在 NBMA（Non Broadcast Multi Access，非广播多路访问）环境下十分有用。在如图 16-4 所示网络中，路由器 2 通过帧中继连接路由器 1 和路由器 3，两个 PVC 都在路由器 2 的同一个物理接口（S0）中止。如果路由器 2 的水平分割功能未被关闭，那么路由器 3 将收不到路由器 1 的路由选择信息（反之亦然）。用 no ip split-horizon 接口子命令可关闭水平分割功能。

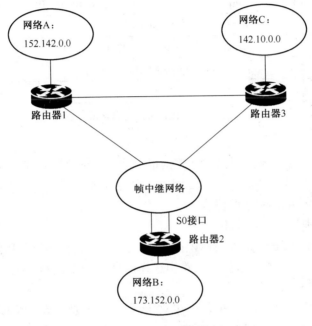

图 16-4　水平分割法原理

3）破坏逆转的水平分割法。

水平分割是路由器用来防止把一个接口得来的路径又从此接口传回而导致问题的方案。水平分割方案忽略在更新过程中从一个路由器获取的路径又传回该路由器。有破坏逆转的水平分割方法是在更新信息中包括这些回传路径，但这种处理方法会把这些回传路径的跳数设为 16（无穷）。通过把跳数设为无穷并把这条路径告诉源路由器，有可能立刻解决路由选择环路。否则，不正确的路径将在路由表中驻留到超时为止。破坏逆转的缺点是它增加了路由更新的数据大小。

4）保持定时器法。

保持定时器法可防止路由器在路径从路由表中删除后一定的时间内（通常为 180 秒）接收新的路由信息。它的作用是保证每个路由器都收到了路径不可达信息，而且没有路由器发出无效路径信息。例如在如图 16-2 所示的网络中，由于路由更新信息被延迟，路由器 2 向路由器 3 发出错误信息。但使用保持定时器法后，这种情况将不会发生，因为路由器 3 将在 180 秒内不接收通向网络 A 的新的路径信息，到那时，路由器 2 将存储正确的路由信息。

5）触发更新法。

有破坏逆转的水平分割将任何两个路由器构成的环路打破，但三个或更多个路由器构成的环路仍会发生,直到无穷(16)时为止。触发更新法可加速收敛时间,它的工作原理是当某个路径的跳数改变了,路由器立即发出更新信息,不管路由器是否到达常规信息更新时间都发出更新信息。

(3) RIP 报文格式。

如图 16-5 所示为 RIP 信息格式。

| 8bits | 16bits | 32bits |
|---|---|---|
| Command | Version | Unused |
| Address Family Identifier | | Route Tag(only for RIP2:0 for RIP) |
| IP Address | | |
| Subnet Mask(only for RIP2:0 for RIP) | | |
| Next hop(only for RIP2:0 for RIP) | | |
| Metric | | |

图 16-5　RIP 协议信息格式

各字段解释如下:

Command:命令字段,8 位,用来指定数据报用途。命令有五种:Request(请求)、Response(响应)、Traceon(启用跟踪标记,自 v2 版本后已经淘汰)、Traceoff(关闭跟踪标记,自 v2 版本后已经淘汰)和 Reserved(保留)。

Version:RIP 版本号字段,16 位。

Address Family Identifier:地址簇标识符字段,24 位。它指出该入口的协议地址类型。由于 RIP2 版本可能使用几种不同协议传送路由选择信息,所以要使用到该字段。IP 协议地址的 Address Family Identifier 为 2。

Route Tag:路由标记字段,32 位,仅在 v2 版本以上需要,第一版本不用,为 0。用于路由器指定属性,必须通过路由器保存和重新广告。路由标志是分离内部和外部 RIP 路由线路的一种常用方法(路由选择域内的网络传送线路),该方法在 EGP 或 IGP 中都有应用。

IP Address:目标 IP 地址字段,IPv4 地址为 32 位。

Subnet Mask:子网掩码字段,IPv4 子网掩码地址为 32 位。它应用于 IP 地址,生成非主机地址部分。如果为 0,说明该入口不包括子网掩码。此字段也仅在 v2 版本以上需要,在 RIPv1 中不需要,为 0。

Next Hop:下一跳字段。指出下一跳 IP 地址,由路由入口指定的通向目的地的数据报需要转发到该地址。

Metric:跳数字段。表示从主机到目的地获得数据报过程中的整个成本。

**四、实验需要掌握的命令**

router(config)#router rip
#启动 RIP 路由进程

router(config-router)#network *network-number*
#将网络和 RIP 进程关联起来

router(config)#neighbor *ip-address*

#给指定的邻居发送单播 RIP 更新

router(config-router)#version {1|2}
#指定要使用的 RIP 版本

router(config-router)#timers basic *update invalid holddown flush* [*sleeptime*]
#调整路由选择协议计时器，*update* 为发送更新路由的时间，*invalid* 为宣告路由无效的时间，*invalid* 至少为 *update* 的三倍，*holddown* 为路由信息被抑制的时间，*holddown* 至少为 *update* 的三倍，*flush* 为将路由从路由表中删除之前等待的时间，*flush* 应该大于 *holddown* 与 *invalid*，*sleeptime* 为 flash 更新时推迟路由更新的时间间隔

router(config-router)#no auto-summary
#禁止路由自动汇总

router(config-router)#auto-summary
#启用路由自动汇总（默认值）

router(config-router)#passive-interface *type number*
#配置接口不发送 RIP 广播和组播更新

router(config-if)#ip rip send version {1|2|1 2}
#改变给定接口的 RIP 发送参数

router(config-if)#ip rip receive version {1|2|1 2}
#改变给定接口的 RIP 接收参数

router(config-if)#no ip split-horizon
#禁止水平分割

router(config-if)#ip split-horizon
#启用水平分割（默认值）

router(config-if)#ip rip authentication key-chain *chain-name*
#启用 RIP 分组的验证

router(config-if)#ip rip authentication mode [md5|test]
#选择验证模式

router(config)#key chain *key-name*
#定义密钥链，一般每个端口使用一个密钥链

router(config-keychain)#key *number*
#配置密钥链中的编号密钥

router(config-keychain-key)#key-string *text*
#定义密钥的文本字符串

router(config-keychain-key)#accept-lifetime *start-time* {infinite|*end-time*|duration *second*}
#指定能够接收有效密钥的时间长度，*start-time* 为开始时间，格式是 hh:mm:ss month date year 或者 hh:mm:ss date month year；*end-time* 为结束时间，格式和 *start-time* 的格式一样；infinite 让密钥自 *start-time* 起被接收；duration 是在启动后持续的时间，单位是秒

router(config-keychain-key)#send-lifetime *start-time* {infinite|*end-time*|duration *second*}
#指定能够发送有效密钥的时间长度，参数和 accept-lifetime 命令的参数一样

router# show ip protocol
#显示 IP 协议的详细信息

router#show ip rip database
#显示 RIP 路由数据库的信息
show ip show ip interface *type number*
#显示指定端口的 IP 详细信息

### 五、实验拓扑图和网络文件

（1）网络拓扑图。

网络拓扑图如图 16-6 所示。

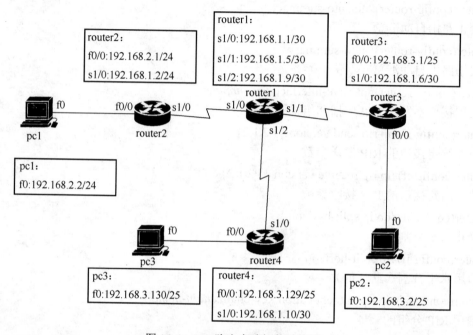

图 16-6　RIP 路由实验拓扑结构图

（2）网络文件。

#RIP.net 文件开始
autostart = false
ghostios = true
sparsemem = true

[localhost]

  [[3660]]
  image = ..\C3660-IS.BIN
  ram = 160

  [[1720]]
  image = ..\C1700-Y.BIN
  ram = 16

  [[ROUTER route1]]

model = 3660
idlepc = 0x60719b98
slot1 = NM-4T
s1/0 = route2 s1/0
s1/1 = route3 s1/0
s1/2 = route4 s1/0

[[ROUTER route2]]
model = 3660
idlepc = 0x60719b98
slot1 = NM-4T
f0/0 = pc1 f0

[[ROUTER route3]]
model = 3660
idlepc = 0x60719b98
slot1 = NM-4T
f0/0 = pc2 f0

[[ROUTER route4]]
model = 3660
idlepc = 0x60719b98
slot1 = NM-4T
f0/0 = pc3 f0

[[router pc1]]
model = 1720
idlepc = 0x8014e4ec

[[router pc2]]
model = 1720
idlepc = 0x8014e4ec

[[router pc3]]
model = 1720
idlepc = 0x8014e4ec
#RIP.net 文件结束

## 六、实验配置步骤

（1）基本配置步骤。
1）router1 的基本配置。
router>enable
router#configure terminal
router(config)#hostname router1
router1(config)#interface Serial1/0
router1(config-if)#ip address 192.168.1.1 255.255.255.252
router1(config-if)#clock rate 9600

```
router1(config-if)#no shutdown
router1(config-if)#interface Serial1/1
router1(config-if)#ip address 192.168.1.5 255.255.255.252
router1(config-if)#clock rate 9600
router1(config-if)#no shutdown
router1(config-if)#interface Serial1/2
router1(config-if)#ip address 192.168.1.9 255.255.255.252
router1(config-if)#clock rate 9600
router1(config-if)#no shutdown
router1(config-if)#end
router1#write
```

2）router2 的基本配置。

```
router>enable
router#configure terminal
router(config)#hostname router2
router2(config)#interface Serial1/0
router2(config-if)#ip address 192.168.1.2 255.255.255.252
router2(config-if)#clock rate 9600
router2(config-if)#no shutdown
router2(config-if)#interface FastEthernet0/0
router2(config-if)#ip address 192.168.2.1 255.255.255.0
router2(config-if)#speed 100
router2(config-if)#duplex full
router2(config-if)#no shutdown
router2(config-if)#end
router2#write
```

3）router3 的基本配置。

```
router>enable
router#configure terminal
router(config)#hostname router3
router3(config)#interface Serial1/0
router3(config-if)#ip address 192.168.1.6 255.255.255.252
router3(config-if)#clock rate 9600
router3(config-if)#no shutdown
router3(config-if)#interface FastEthernet0/0
router3(config-if)#ip address 192.168.3.1 255.255.255.128
router3(config-if)#speed 100
router3(config-if)#duplex full
router3(config-if)#no shutdown
router3(config-if)#end
router3#write
```

4）router4 的基本配置。

```
router>enable
router#configure terminal
router(config)#hostname router4
router4(config)#interface Serial1/0
router4(config-if)#ip address 192.168.1.10 255.255.255.252
```

router4(config-if)#clock rate 9600
router4(config-if)#no shutdown
router4(config-if)#interface FastEthernet0/0
router4(config-if)#ip address 192.168.3.129 255.255.255.128
router4(config-if)#speed 100
router4(config-if)#duplex full
router4(config-if)#no shutdown
router4(config-if)#end
router4#write

5）主机 pc1 的配置步骤。

router>enable
router#configure terminal
router(config)#hostname pc1
pc1(config)#interface FastEthernet0
pc1(config-if)#ip address 192.168.2.2 255.255.255.0
pc1(config-if)#duplex full
pc1(config-if)#no shutdown
pc1(config-if)#exit
pc1(config)#no ip routing
pc1(config)#ip default-gateway 192.168.2.1
pc2(config)#exit
pc2#write

6）主机 pc2 的配置步骤。

router>enable
router#configure terminal
router(config)#hostname pc2
pc2(config)#interface FastEthernet0
pc2(config-if)#ip address 192.168.3.2 255.255.255.128
pc2(config-if)#duplex full
pc2(config-if)#no shutdown
pc2(config-if)#exit
pc2(config)#no ip routing
pc2(config)#ip default-gateway 192.168.3.1
pc2(config)#exit
pc2#write

7）主机 pc3 的配置步骤。

router>enable
router#configure terminal
router(config)#hostname pc3
pc3(config)#interface FastEthernet0
pc3(config-if)#ip address 192.168.3.130 255.255.255.128
pc3(config-if)#duplex full
pc3(config-if)#no shutdown
pc3(config-if)#exit
pc3(config)#no ip routing
pc3(config)#ip default-gateway 192.168.3.129
pc3(config)#exit

pc3#write

（2）RIP 的配置。

1）路由器 router1 的配置。

#下面是配置 RIP 验证密钥链

router1(config)#key chain router2
router1(config-keychain)#key 1
router1(config-keychain-key)#key-string cisco
router1(config-keychain-key)#exit
router1(config-keychain)#exit
router1(config)#key chain router3
router1(config-keychain)#key 1
router1(config-keychain-key)#key-string cisco
router1(config-keychain-key)#exit
router1(config-keychain)#exit
router1(config)#key chain router4
router1(config-keychain)#key 1
router1(config-keychain-key)#key-string cisco
router1(config-keychain-key)#exit
router1(config-keychain)#exit

#下面是将三个验证密钥链分别应用到三个端口上，两个对接的端口的密钥链配置和验证模式应完全一样，否则无法接收更新路由

router1(config)#interface Serial1/0
router1(config-if)#ip rip authentication mode md5
router1(config-if)#ip rip authentication key-chain router2
router1(config-if)#interface Serial1/1
router1(config-if)#ip rip authentication mode md5
router1(config-if)#ip rip authentication key-chain router3
router1(config-if)#interface Serial1/2
router1(config-if)#ip rip authentication mode md5
router1(config-if)#ip rip authentication key-chain router4
router1(config-if)#exit

#下面是启动 RIP 进程、计时器、禁止路由汇总、宣告网络和协议的运行版本等配置

router1(config)#router rip
router1(config-router)#version 2
router1(config-router)#timers basic 20 80 80 120 1000
router1(config-router)#network 192.168.1.0
router1(config-router)#no auto-summary

2）路由器 router2 的配置。

#下面是配置 RIP 验证密钥链

router2(config)#key chain router2
router2(config-keychain)#key 1
router2(config-keychain-key)#key-string cisco
router2(config-keychain-key)#exit
router2(config-keychain)#exit

#下面是将验证密钥链应用到 s1/0 端口上，与 router1 的 s1/0 端口的密钥链配置和验证模式应完全一样，否则无法接收更新路由

router2(config)#interface Serial1/0

```
router2(config-if)#ip rip authentication mode md5
router2(config-if)#ip rip authentication key-chain router2
#下面是启动 RIP 进程、计时器、宣告网络和协议的运行版本等配置
router2(config)#router rip
router2(config-router)#version 2
router2(config-router)#timers basic 20 80 80 120 1000
router2(config-router)#network 192.168.1.0
router2(config-router)#network 192.168.2.0
router2(config-router)#no auto-summary
```

3）路由器 router3 的配置。

```
#下面是配置 RIP 验证密钥链
router3(config)#key chain router3
router3(config-keychain)#key 1
router3(config-keychain-key)#key-string cisco
router3(config-keychain-key)#exit
router3(config-keychain)#exit
#下面是将验证密钥链应用到 s1/0 端口上，与 router1 的 s1/1 端口的密钥链配置和验证模式应完全一样，否则无法接收更新路由
router3(config)#interface Serial1/0
router3(config-if)#ip rip authentication mode md5
router3(config-if)#ip rip authentication key-chain router3
#下面是启动 RIP 进程、计时器、禁止路由汇总、宣告网络和协议的运行版本等配置
router3(config)#router rip
router3(config-router)#version 2
router3(config-router)#timers basic 20 80 80 120 1000
router3(config-router)#network 192.168.1.4
router3(config-router)#network 192.168.3.0
router3(config-router)#no auto-summary
```

4）路由器 router4 的配置。

```
#下面是配置 RIP 验证密钥链
router4(config)#key chain router4
router4(config-keychain)#key 1
router4(config-keychain-key)#key-string cisco
router4(config-keychain-key)#exit
router4(config-keychain)#exit
#下面是将验证密钥链应用到 s1/0 端口上，与 router1 的 s1/2 端口的密钥链配置和验证模式应完全一样，否则无法接收更新路由
router4(config)#interface Serial1/0
router4(config-if)#ip rip authentication mode md5
router4(config-if)#ip rip authentication key-chain router4
#下面是启动 RIP 进程、计时器、禁止路由汇总、宣告网络和协议的运行版本等配置
router4(config)#router rip
router4(config-router)#version 2
router4(config-router)#timers basic 20 80 80 120 1000
router4(config-router)#network 192.168.1.8
router4(config-router)#network 192.168.3.0
router4(config-router)#no auto-summary
```

此处省略了 pc1、pc2 和 pc3 的配置信息，参照前面的实验进行配置。

## 七、实验测试和查看相关配置

（1）测试基本配置。

测试每一对直连端口通信是否正常，例如：在主机 pc1 上 ping 路由器 router2 的 f0/0 端口，结果如下，说明这一对直连端口通信正常。如果没有收到任何 ICMP 的应答数据包，则说明这一对直连端口通信不正常，这时要查看端口的 IP 地址、子网掩码、speed、duplex 等参数的配置。

pc1>ping 192.168.2.1

Type escape sequence to abort.
Sending 5, 100-byte ICMP Echos to 192.168.2.1, timeout is 2 seconds:
!!!!!
Success rate is 100 percent (5/5), round-trip min/avg/max = 152/202/260 ms

（2）RIP 的配置。

R  192.168.2.0/24 [120/1] via 192.168.1.2, 00:00:12, Serial1/0 ：路由中 R 的意思是此条路由是由 RIP 路由协议生成的动态路由；192.168.2.0/24，其中 192.168.2.0 是目的网段地址，/24 的意思是目的网段子网掩码前缀为 24 位；[1/0]的意思是管理距离为 1，度量值为 0；Serial1/0 的意思是下一跳的接口为 Serial1/0。本条路由的意思是要到达目的网段 192.168.2.0/24 子网需要通过 Serial1/0 接口进行转发。

192.168.1.0/30 is subnetted, 3 subnets 的意思是在 192.168.10 此网段有两个子网，子网掩码为 30 位。

C  192.168.1.8 is directly connected, Serial1/2：路由中 C 代表此条路由是直连路由；本条路由的意思是 192.168.1.8/30 子网和路由器直接相连，直接相连的接口是 Serial1/2。

Gateway of last resort is not set

router1(config-router)#do show ip route
Codes: C - connected, S - static, R - RIP, M - mobile, B - BGP
   D - EIGRP, EX - EIGRP external, O - OSPF, IA - OSPF inter area
   N1 - OSPF NSSA external type 1, N2 - OSPF NSSA external type 2
   E1 - OSPF external type 1, E2 - OSPF external type 2
   i - IS-IS, su - IS-IS summary, L1 - IS-IS level-1, L2 - IS-IS level-2
   ia - IS-IS inter area, * - candidate default, U - per-user static route
   o - ODR, P - periodic downloaded static route

   192.168.1.0/30 is subnetted, 3 subnets
C  192.168.1.8 is directly connected, Serial1/2
C  192.168.1.0 is directly connected, Serial1/0
C  192.168.1.4 is directly connected, Serial1/1
R  192.168.2.0/24 [120/1] via 192.168.1.2, 00:00:12, Serial1/0
   192.168.3.0/25 is subnetted, 2 subnets
R  192.168.3.0 [120/1] via 192.168.1.6, 00:00:17, Serial1/1
R  192.168.3.128 [120/1] via 192.168.1.10, 00:00:07, Serial1/2

router2#show ip route
Codes: C - connected, S - static, R - RIP, M - mobile, B - BGP
   D - EIGRP, EX - EIGRP external, O - OSPF, IA - OSPF inter area
   N1 - OSPF NSSA external type 1, N2 - OSPF NSSA external type 2
   E1 - OSPF external type 1, E2 - OSPF external type 2

         i - IS-IS, su - IS-IS summary, L1 - IS-IS level-1, L2 - IS-IS level-2
         ia - IS-IS inter area, * - candidate default, U - per-user static route
         o - ODR, P - periodic downloaded static route

Gateway of last resort is not set

         192.168.1.0/30 is subnetted, 3 subnets
R        192.168.1.8 [120/1] via 192.168.1.1, 00:00:08, Serial1/0
C        192.168.1.0 is directly connected, Serial1/0
R        192.168.1.4 [120/1] via 192.168.1.1, 00:00:08, Serial1/0
C     192.168.2.0/24 is directly connected, FastEthernet0/0
         192.168.3.0/25 is subnetted, 2 subnets
R        192.168.3.0 [120/2] via 192.168.1.1, 00:00:08, Serial1/0
R        192.168.3.128 [120/2] via 192.168.1.1, 00:00:08, Serial1/0

router3(config-router)#do sho ip route
Codes: C - connected, S - static, R - RIP, M - mobile, B - BGP
         D - EIGRP, EX - EIGRP external, O - OSPF, IA - OSPF inter area
         N1 - OSPF NSSA external type 1, N2 - OSPF NSSA external type 2
         E1 - OSPF external type 1, E2 - OSPF external type 2
         i - IS-IS, su - IS-IS summary, L1 - IS-IS level-1, L2 - IS-IS level-2
         ia - IS-IS inter area, * - candidate default, U - per-user static route
         o - ODR, P - periodic downloaded static route

Gateway of last resort is not set

         192.168.1.0/30 is subnetted, 3 subnets
R        192.168.1.8 [120/1] via 192.168.1.5, 00:00:17, Serial1/0
R        192.168.1.0 [120/1] via 192.168.1.5, 00:00:17, Serial1/0
C        192.168.1.4 is directly connected, Serial1/0
R     192.168.2.0/24 [120/2] via 192.168.1.5, 00:00:17, Serial1/0
         192.168.3.0/25 is subnetted, 2 subnets
C        192.168.3.0 is directly connected, FastEthernet0/0
R        192.168.3.128 [120/2] via 192.168.1.5, 00:00:17, Serial1/0

router4#show ip route
Codes: C - connected, S - static, R - RIP, M - mobile, B - BGP
         D - EIGRP, EX - EIGRP external, O - OSPF, IA - OSPF inter area
         N1 - OSPF NSSA external type 1, N2 - OSPF NSSA external type 2
         E1 - OSPF external type 1, E2 - OSPF external type 2
         i - IS-IS, su - IS-IS summary, L1 - IS-IS level-1, L2 - IS-IS level-2
         ia - IS-IS inter area, * - candidate default, U - per-user static route
         o - ODR, P - periodic downloaded static route

Gateway of last resort is not set

         192.168.1.0/30 is subnetted, 3 subnets
C        192.168.1.8 is directly connected, Serial1/0

R         192.168.1.0 [120/1] via 192.168.1.9, 00:00:09, Serial1/0
R         192.168.1.4 [120/1] via 192.168.1.9, 00:00:09, Serial1/0
R      192.168.2.0/24 [120/2] via 192.168.1.9, 00:00:09, Serial1/0
       192.168.3.0/25 is subnetted, 2 subnets
R         192.168.3.0 [120/2] via 192.168.1.9, 00:00:09, Serial1/0
C         192.168.3.128 is directly connected, FastEthernet0/0

如果路由器 router3 和 router4 不禁止路由汇总，将会导致路由出错。结果如下：
router1#show ip route
Codes: C - connected, S - static, R - RIP, M - mobile, B - BGP
       D - EIGRP, EX - EIGRP external, O - OSPF, IA - OSPF inter area
       N1 - OSPF NSSA external type 1, N2 - OSPF NSSA external type 2
       E1 - OSPF external type 1, E2 - OSPF external type 2
       i - IS-IS, su - IS-IS summary, L1 - IS-IS level-1, L2 - IS-IS level-2
       ia - IS-IS inter area, * - candidate default, U - per-user static route
       o - ODR, P - periodic downloaded static route

Gateway of last resort is not set

       192.168.1.0/30 is subnetted, 3 subnets
C         192.168.1.8 is directly connected, Serial1/2
C         192.168.1.0 is directly connected, Serial1/0
C         192.168.1.4 is directly connected, Serial1/1
R      192.168.2.0/24 [120/1] via 192.168.1.2, 00:00:10, Serial1/0
       192.168.3.0/24 is variably subnetted, 2 subnets, 2 masks
R         192.168.3.0/25 [120/1] via 192.168.1.6, 00:00:43, Serial1/1
R         192.168.3.0/24 [120/1] via 192.168.1.10, 00:00:03, Serial1/2
                        [120/1] via 192.168.1.6, 00:00:04, Serial1/1 ＃两条不同的路径到达
192.168.3.0/24 的网络，与实际网络不符，实际的网络将 192.168.3.0/24 划分为两个子网，192.168.3.0/25 的子网通过 s1/1 端口转发，而 192.168.3.128/25 的子网通过 s1/2 转发。

router3(config-router)#do show ip route
Codes: C - connected, S - static, R - RIP, M - mobile, B - BGP
       D - EIGRP, EX - EIGRP external, O - OSPF, IA - OSPF inter area
       N1 - OSPF NSSA external type 1, N2 - OSPF NSSA external type 2
       E1 - OSPF external type 1, E2 - OSPF external type 2
       i - IS-IS, su - IS-IS summary, L1 - IS-IS level-1, L2 - IS-IS level-2
       ia - IS-IS inter area, * - candidate default, U - per-user static route
       o - ODR, P - periodic downloaded static route

Gateway of last resort is not set

       192.168.1.0/30 is subnetted, 3 subnets
R         192.168.1.8 [120/1] via 192.168.1.5, 00:00:10, Serial1/0
R         192.168.1.0 [120/1] via 192.168.1.5, 00:00:10, Serial1/0
C         192.168.1.4 is directly connected, Serial1/0
R      192.168.2.0/24 [120/1] via 192.168.1.5, 00:00:10, Serial1/0
       192.168.3.0/25 is subnetted, 1 subnets

C        192.168.3.0 is directly connected, FastEthernet0/0   #没有到达 192.168.3.128/25 的子网的路由

（3）测试网络。

下面是在主机 pc1 上 ping 主机 pc2 和 pc3，通信正常，说明网络配置没有问题。

pc1>ping 192.168.3.2

Type escape sequence to abort.
Sending 5, 100-byte ICMP Echos to 192.168.3.2, timeout is 2 seconds:
!!!!!
Success rate is 100 percent (5/5), round-trip min/avg/max = 152/202/260 ms
pc1>ping 192.168.3.129

Type escape sequence to abort.
Sending 5, 100-byte ICMP Echos to 192.168.3.129, timeout is 2 seconds:
!!!!!
Success rate is 100 percent (5/5), round-trip min/avg/max = 72/165/300 ms

（4）查看 RIP 路由的版本和计时器的数值。

router1(config-router)# do show ip protocol
Routing Protocol is "rip"
　Outgoing update filter list for all interfaces is not set    #输出更新过滤列表没有设置
　Incoming update filter list for all interfaces is not set    #输入更新过滤列表没有设置
　Sending updates every 20 seconds, next due in 9 seconds    #每 20 秒钟发送更新路由信息，下一个更新周期还有 9 秒钟
　Invalid after 80 seconds, hold down 80, flushed after 120
　Redistributing: rip    #下面显示路由重定向的信息，此例中没有相关的信息
　Default version control: send version 2, receive version 2    #显示的是每个端口发送和接收 RIP 路由信息的版本以及认真的密钥链

| Interface | Send | Recv | Triggered RIP | Key-chain |
|---|---|---|---|---|
| Serial1/0 | 2 | 2 |  | router2 |
| Serial1/1 | 2 | 2 |  | router3 |
| Serial1/2 | 2 | 2 |  | router4 |

　Automatic network summarization is not in effect    #路由自动汇总没有设置
　Maximum path: 4    #路径数量最大为 4
　Routing for Networks:    #直连的网络
　　192.168.1.0
　Routing Information Sources:    #直连的网络的网关、距离、最后一次更新的时间

| Gateway | Distance | Last Update |
|---|---|---|
| 192.168.1.10 | 120 | 00:00:06 |
| 192.168.1.2 | 120 | 00:00:18 |
| 192.168.1.6 | 120 | 00:00:00 |

　Distance: (default is 120)

（5）查看路由器交换路由信息。

router1#debug ip rip database
RIP database events debugging is on
router1#show debug
Oct  1 05:56:21.890: RIP-DB: network_update with 192.168.3.0/25 succeeds    #192.168.3.0/25 网络路由更新成功
Oct  1 05:56:21.890: RIP-DB: adding 192.168.3.0/25 (metric 1) via 192.168.1.6 on Serial1/1 to RIP database

#增加到网络 192.168.3.0/25 的路由，此路由的下一跳为 192.168.1.6，来源于 Serial1/1 接口
Oct　1 05:56:35.482: RIP-DB: network_update with 192.168.2.0/24 succeeds
Oct　1 05:56:35.486: RIP-DB: adding 192.168.2.0/24 (metric 1) via 192.168.1.2 on Serial1/0 to RIP database
Oct　1 05:56:35.718: RIP-DB: network_update with 192.168.3.128/25 succeeds
Oct　1 05:56:35.718: RIP-DB: adding 192.168.3.128/25 (metric 1) via 192.168.1.10 on Serial1/2 to RIP database
Oct　1 05:56:41.862: RIP-DB: network_update with 192.168.3.0/25 succeeds
Oct　1 05:56:41.862: RIP-DB: adding 192.168.3.0/25 (metric 1) via 192.168.1.6 on Serial1/1 to RIP database

（6）查看端口的水平分割是否启用。

router1(config-router)#do show ip interface s1/0
Serial1/0 is up, line protocol is up
　　Internet address is 192.168.1.1/30
　　Broadcast address is 255.255.255.255
　　Address determined by setup command
　　MTU is 1500 bytes
　　Helper address is not set
　　Directed broadcast forwarding is disabled
　　Multicast reserved groups joined: 224.0.0.9
　　Outgoing access list is not set
　　Inbound　access list is not set
　　Proxy ARP is enabled
　　Local Proxy ARP is disabled
　　Security level is default
　　**Split horizon is enabled**　　#端口启用了水平分割功能
　　……

（7）测试水平分割功能。

如果在路由器 router2 s1/0 端口模式下用 no ip split-horizon 命令关闭水平分割功能，则会出现路由器 router1 上的路由异常，结果如下：

router1#sho ip route
Codes: C - connected, S - static, R - RIP, M - mobile, B - BGP
　　　　D - EIGRP, EX - EIGRP external, O - OSPF, IA - OSPF inter area
　　　　N1 - OSPF NSSA external type 1, N2 - OSPF NSSA external type 2
　　　　E1 - OSPF external type 1, E2 - OSPF external type 2
　　　　i - IS-IS, su - IS-IS summary, L1 - IS-IS level-1, L2 - IS-IS level-2
　　　　ia - IS-IS inter area, * - candidate default, U - per-user static route
　　　　o - ODR, P - periodic downloaded static route

Gateway of last resort is not set

　　　　192.168.1.0/30 is subnetted, 3 subnets
C　　　　192.168.1.8 is directly connected, Serial1/2
C　　　　192.168.1.0 is directly connected, Serial1/0
C　　　　192.168.1.4 is directly connected, Serial1/1
R　　　　192.168.2.0/24 [120/1] via 192.168.1.2, 00:00:04, Serial1/0
　　　　192.168.3.0/24 is variably subnetted, 3 subnets, 2 masks
R　　　　192.168.3.0/25 [120/1] via 192.168.1.6, 00:00:10, Serial1/1
R　　　　192.168.3.0/24 [120/3] via 192.168.1.2, 00:00:04, Serial1/0　　#此条路由是开启水平分割功能之后

生成的路由，此条路由有问题

R        192.168.3.128/25 [120/1] via 192.168.1.10, 00:00:09, Serial1/2

## 八、注意问题

（1）两个直接连接的 Serial 端口的时钟速率一定要一致。

（2）两个直接连接的 Serial 端口封装的协议要一致，本实验都是 HDLC（默认值）。

（3）配置 RIP 动态路由验证时，要保持验证两端的密钥链和模式一致。

（4）在配置路由汇总时，要保证动态生成的路由正确，否则要关闭路由汇总，主要问题是划分的子网掩码不是 A、B、C 类的默认子网掩码。

（5）要注意启用水平分割功能，此功能默认是开启的。

（6）修改使用 RIP 动态路由版本号时，注意端口设置的版本优先于进程下设置的版本号。

（7）在调整路由选择协议计时器时，要保持所有运行路由器的计时器一致。

# 实验十七  OSPF 路由协议的配置

OSPF 路由协议是一种典型的链路状态（Link-state）路由协议，一般用于同一个路由域内。在这里，路由域是指一个自治系统（Autonomous System，AS），它是指一组通过统一的路由政策或路由协议互相交换路由信息的网络。在这个 AS 中，所有的 OSPF 路由器都维护一个相同的描述这个 AS 结构的数据库，该数据库中存放的是路由域中相应链路的状态信息，OSPF 路由器正是通过这个数据库计算出其 OSPF 路由表的。OSPF 将链路状态广播数据包 LSA（Link State Advertisement）传送给在某一区域内的所有路由器，这一点与距离矢量路由协议不同。运行距离矢量路由协议的路由器是将部分或全部的路由表传递给与其相邻的路由器。

RIP 的 15 跳限制；RIP 不能支持可变长子网掩码（VLSM）；周期性广播整个路由表，收敛速度慢；RIP 路由选路基于跳数；RIP 没有区域的概念。相对 RIP 而言，OSPF 更适合用于大型网络：没有跳数的限制；支持可变长子网掩码（VLSM）；使用组播发送链路状态更新；在链路状态变化时使用触发更新；收敛速度快。

## 一、实验目的

（1）掌握 OSPF 的工作原理；
（2）熟练掌握 OSPF 的各种配置。

## 二、实验内容

（1）OSPF 的基本配置；
（2）多区域的 OSPF 协议配置；
（3）虚链路的配置；
（4）指定路由器（DR）的配置；
（5）OSPF 路由交换的安全配置；
（6）OSPF 定时器的配置；
（7）禁止路由汇总的配置。

要求：

（1）在 Area 1 中，路由器 R4、R5 和 R6 通过一台交换机构成的广播局域网络互连，各路由器 ID 由路由器的 loopback 接口地址指定，如指定 R4 是指派路由器（Designated Routers，DR）、R5 为备份的指派路由器（Backup Designated Router，BDR），而 R6 不参与指派路由器的选择过程；

（2）要求网络全连通。

## 三、OSPF 协议的简介

OSPF 路由协议是一种典型的链路状态（Link-state）路由协议，一般用于同一个路由域内。在这里，路由域是指一个自治系统（Autonomous System，AS），它是指一组通过统一的路由政策或路由协议互相交换路由信息的网络。在这个 AS 中，所有的 OSPF 路由器都维护一个相同的描述这个 AS 结构的数据库，该数据库中存放的是路由域中相应链路的状态信息，OSPF 路由器正是通过这个数据库计算出其 OSPF 路由表的。

作为一种链路状态的路由协议，OSPF 将链路状态广播数据包 LSA（Link State Advertisement）传送给在某一区域内的所有路由器，这一点与距离矢量路由协议不同。运行距离矢量路由协议的路由器是将部分或全部的路由表传递给与其相邻的路由器。

（1）数据包格式。

OSPF 路由协议的数据包格式如图 17-1 所示。

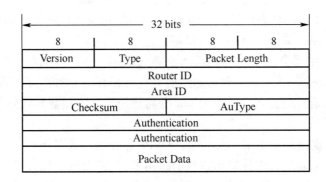

图 17-1　OSPF 路由协议的数据包格式

在 OSPF 路由协议的数据包中，其数据包头长为 24 个字节，包含如下 8 个字段：

1）Version：定义所采用的 OSPF 路由协议的版本。

2）Type：定义 OSPF 数据包类型。OSPF 数据包共有以下五种：

① Hello：用于建立和维护相邻的两个 OSPF 路由器的关系，该数据包是周期性发送的。

② Database Description：用于描述整个数据库，该数据包仅在 OSPF 初始化时发送。

③ Link state request：用于向相邻的 OSPF 路由器请求部分或全部的数据，这种数据包是在当路由器发现其数据已经过期时才发送的。

④ Link State Update：这是对 Link State 请求数据包的响应，即通常所说的 LSA 数据包。

⑤ Link State Acknowledgment：是对 LSA 数据包的响应。

3）Packet Length：定义整个数据包的长度。

4）Router ID：用于描述数据包的源地址，以 IP 地址来表示。

5）Area ID：用于区分 OSPF 数据包属于的区域号，所有的 OSPF 数据包都属于一个特定的 OSPF 区域。

6）Checksum：校验位，用于标记数据包在传递时有无误码。

7）AuType：定义 OSPF 验证类型。

8）Authentication：包含 OSPF 验证信息，长为 8 个字节。

（2）OSPF 基本算法。

1）SPF 算法及最短路径树。

SPF 算法是 OSPF 路由协议的基础。SPF 算法有时也被称为 Dijkstra 算法，这是因为最短路径优先算法（SPF）是 Dijkstra 发明的。SPF 算法将每一个路由器作为根（ROOT）来计算其到每一个目的地路由器的距离，每一个路由器根据一个统一的数据库计算出路由域的拓扑结构图，该结构图类似于一棵树，在 SPF 算法中被称为最短路径树。在 OSPF 路由协议中，最短路径树的树干长度（即 OSPF 路由器至每一个目的地路由器的距离）称为 OSPF 的 Cost，其算法为：Cost = $100 \times 10^6$/链路带宽，在这里，链路带宽以 b/s 来表示。也就是说，OSPF 的 Cost 与链路的带宽成反比，带宽越高，Cost 越小，表示 OSPF 到目的地的距离越近。举例来说，FDDI 或快速以太网的 Cost 为 1，2M 串行链路的 Cost 为 50，10M 以太网的 Cost 为 10 等。

2）链路状态算法。

作为一种典型的链路状态的路由协议，OSPF 还得遵循链路状态路由协议的统一算法。链路状态的算法非常简单，这里将链路状态算法概括为以下四个步骤：

第一步：当路由器初始化或当网络结构发生变化（例如增减路由器、链路状态发生变化等）时，路由器会产生链路状态广播数据包 LSA（Link State Advertisement），该数据包里包含路由器上所有相连链路，即为所有端口的状态信息。

第二步：所有路由器会通过一种被称为刷新（Flooding）的方法来交换链路状态数据。Flooding 是指路由器将其 LSA 数据包传送给所有与其相邻的 OSPF 路由器，相邻路由器根据其接收到的链路状态信息更新自己的数据库，并将该链路状态信息转送给与其相邻的路由器，直至稳定的一个过程。

第三步：当网络重新稳定下来，也可以说 OSPF 路由协议收敛下来时，所有的路由器会根据其各自的链路状态信息数据库计算出各自的路由表。该路由表中包含路由器到每一个可到达目的地的 Cost 以及到达该目的地所要转发的下一个路由器（Next Hop）。

第四步：当网络状态比较稳定时，网络中传递的链路状态信息是比较少的，或者可以说，当网络稳定时，网络中是比较安静的。这也正是链路状态路由协议区别于距离矢量路由协议的一大特点。

（3）OSPF 的基本特征。

OSPF 路由协议是一种链路状态的路由协议，为了更好地说明 OSPF 路由协议的基本特征，我们将 OSPF 路由协议与距离矢量路由协议中的 RIP（Routing Information Protocol）作比较，归纳为如下几点：

RIP 路由协议中用于表示目的网络远近的唯一参数为跳（Hop），即到达目的网络所要经过的路由器个数。在 RIP 路由协议中，该参数被限制为最大 15，也就是说 RIP 路由信息最多能传递至第 16 个路由器；对于 OSPF 路由协议，路由表中表示目的网络的参数为 Cost，该参数为一虚拟值，与网络中链路的带宽等相关，也就是说 OSPF 路由信息不受物理跳数的限制。并且，OSPF 路由协议还支持 ToS（Type of Service）路由，因此，OSPF 比较适合应用于大型网络中。

RIP 路由协议不支持变长子网屏掩码（VLSM），这被认为是 RIP 路由协议不适用于大型网络的又一重要原因。采用变长子网屏蔽码可以在最大限度上节约 IP 地址。OSPF 路由协议对 VLSM 有良好的支持性。

RIP 路由协议路由收敛较慢。RIP 路由协议周期性地将整个路由表作为路由信息广播至网络中，该广播周期为 30 秒。在一个较为大型的网络中，RIP 协议会产生很大的广播信息，占用较多的网络带宽资源；并且由于 RIP 协议 30 秒的广播周期，影响了 RIP 路由协议的收敛，甚至出现不收敛的现象。而 OSPF 是一种链路状态的路由协议，当网络比较稳定时，网络中的路由信息是比较少的，并且其广播也不是周期性的，因此 OSPF 路由协议即使是在大型网络中也能够较快地收敛。

在 RIP 协议中，网络是一个平面的概念，并无区域及边界等的定义。随着无级路由 CIDR 概念的出现，RIP 协议就明显落伍了。在 OSPF 路由协议中，一个网络或者说是一个路由域可以划分为很多个区域，每一个区域通过 OSPF 边界路由器相连，区域间可以通过路由汇总（Summary）来减少路由信息，减小路由表，提高路由器的运算速度。

OSPF 路由协议支持路由验证，只有互相通过路由验证的路由器之间才能交换路由信息。并且 OSPF 可以对不同的区域定义不同的验证方式，提高网络的安全性。

OSPF 路由协议对负载分担的支持性能较好。OSPF 路由协议支持多条 Cost 相同的链路上的负载分担，目前一些厂家的路由器支持 6 条链路的负载分担。

（4）区域及域间路由。

在 OSPF 路由协议的定义中，可以将一个路由域或者一个自治系统（AS）划分为几个区域。在 OSPF 中，由按照一定的 OSPF 路由法则组合在一起的一组网络或路由器的集合称为区域（area）。

在 OSPF 路由协议中，每一个区域中的路由器都按照该区域中定义的链路状态算法来计算网络拓扑结构，这意味着每一个区域都有着该区域独立的网络拓扑数据库及网络拓扑图。对于每一个区域，其网络拓扑结构在区域外是不可见的，同样，在每一个区域中的路由器对其域外的其余网络结构也不了解。这意味着 OSPF 路由域中的网络链路状态数据广播被区域的边界挡住了，这样做有利于减少网络中链路状态数据包在全网范围内的广播，也是 OSPF 将其路由域或一个 AS 划分成很多个区域的重要原因。

随着区域概念的引入，不再是在同一个 AS 内的所有路由器都有一个相同的链路状态数据库，而是路由器具有与其相连的每一个区域的链路状态信息，即该区域的结构数据库，当一个路由器与多个区域相连时，我们称之为区域边界路由器。一个区域边界路由器有自身相连的所有区域的网络结构数据。在同一个区域中的两个路由器有着对该区域相同的结构数据库。

我们可以根据 IP 数据包的目的地址及源地址将 OSPF 路由域中的路由分成两类，当目的地址与源地址处于同一个区域中时，称为区域内路由，当目的地址与源地址处于不同的区域甚至处于不同的 AS 时，我们称之为域间路由。

OSPF 的骨干区域及虚拟链路（Virtual Link）：在 OSPF 路由协议中存在一个骨干区域（Backbone），该区域包括属于这个区域的网络及相应的路由器，骨干区域必须是连续的，同时也要求其余区域必须与骨干区域直接相连。骨干区域一般为区域 0，其主要工作是在其余区域间传递路由信息。所有的区域，包括骨干区域之间的网络结构情况是互不可见的，当一个区域的路由信息对外广播时，其路由信息是先传递至区域 0（骨干区域），再由区域 0 将该路由信息向其余区域作广播。

在实际网络中，可能会存在 Backbone 不连续或者某一个区域与骨干区域物理不相连的情况，在这两种情况下，系统管理员可以通过设置虚拟链路的方法来解决。

虚拟链路设置在两个路由器之间，这两个路由器都有一个端口与同一个非骨干区域相连。虚拟链路被认为是属于骨干区域的，在 OSPF 路由协议看来，虚拟链路两端的两个路由器被一个点对点的链路连在一起。在 OSPF 路由协议中，通过虚拟链路的路由信息是作为域内路由来看待的。下面我们分两种情况来说明虚拟链路在 OSPF 路由协议中的作用。

1）当一个区域与 area0 没有物理链路相连时。

一个骨干区域 area 0 必须位于所有区域的中心，其余所有区域必须与骨干区域直接相连。但是，也存在一个区域无法与骨干区域建立物理链路的可能性，在这种情况下，我们可以采用虚拟链路。虚拟链路在该区域与骨干区域间建立一个逻辑连接点,该虚拟链路必须建立在两个区域边界路由器之间,并且其中一个区域边界路由器必须属于骨干区域。这种虚拟链路可以用图 17-2 来说明。

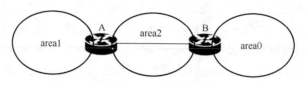

图 17-2  虚拟链路 1

在图 17-2 所示的例子中,区域 1 与区域 0 并无物理相连链路,我们可以在路由器 A 及路由器 B 之间建立虚拟链路,这样,让区域 2 作为一个穿透网络(Transit Network),路由器 B 作为接入点,区域 1 就与区域 0 建立了逻辑联接。

2)当骨干区域不连续时。

OSPF 路由协议要求骨干区域 area0 必须是连续的,但是,骨干区域也会出现不连续的情况,例如,当我们想把两个 OSPF 路由域混合到一起,并且想要使用一个骨干区域时,或者当某些路由器出现故障引起骨干区域不连续,在这些情况下,我们可以采用虚拟链路将两个不连续的区域 0 连接到一起。这时,虚拟链路的两端必须是两个区域 0 的边界路由器,并且这两个路由器必须都有处于同一个区域的端口,以图 17-3 为例。

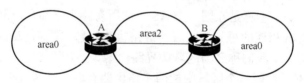

图 17-3  虚拟链路 2

在图 17-3 所示的例子中,穿过区域 2 的虚拟链路将两个分为两半的骨干区域连接到一起,路由器 A 与 B 之间的路由信息作为 OSPF 域内路由来处理。

另外,当一个非骨干区域的区域分裂成两半时,不能采用虚拟链路的方法来解决。当出现这种情况时,分裂出的其中一个区域将被其余的区域作为域间路由来处理。

残域(Stub Area):在 OSPF 路由协议的链路状态数据库中,可以包括 AS 外部链路状态信息,这些信息会通过 Flooding 传递到 AS 内的所有 OSPF 路由器上。但是,在 OSPF 路由协议中存在这样一种区域,我们把它称为残域(Stub Area),AS 外部信息不允许广播进/出这个区域。对于残域来说,访问 AS 外部的数据只能根据默认路由(Default Route)来寻址。这样做有利于减小残域内部路由器上的链路状态数据库的大小及存储器的使用,提高路由器计算路由表的速度。

当一个 OSPF 的区域只存在一个区域出口点时,我们可以将该区域配置成一个残域,此时,该区域的边界路由器会对域内广播默认路由信息。需要注意的是,一个残域中的所有路由器都必须知道自身属于该残域,否则残域的设置没有作用。另外,针对残域还有两点需要注意:一是残域中不允许存在虚拟链路;二是残域中不允许存在 AS 边界路由器。

(5)OSPF 协议数据包分类。

1)OSPF 路由器的分类。

当一个 AS 划分成几个 OSPF 区域时,根据一个路由器在相应的区域中的作用,可以将

OSPF 路由器作如下分类：

①内部路由器：当一个 OSPF 路由器上所有直连的链路都处于同一个区域时，我们称这种路由器为内部路由器。内部路由器上仅仅运行其所属区域的 OSPF 运算法则。

②区域边界路由器：当一个路由器与多个区域相连时，我们称之为区域边界路由器。区域边界路由器运行与其相连的所有区域定义的 OSPF 运算法则，具有相连的每一个区域的网络结构数据，并且了解如何将该区域的链路状态信息广播至骨干区域，再由骨干区域转发至其余区域。

③AS 边界路由器：AS 边界路由器是与 AS 外部的路由器互相交换路由信息的 OSPF 路由器，该路由器在 AS 内部广播其所得到的 AS 外部路由信息；这样 AS 内部的所有路由器都知道至 AS 边界路由器的路由信息。AS 边界路由器的定义是与前面几种路由器的定义相独立的，一个 AS 边界路由器可以是一个区域内部路由器或是一个区域边界路由器。

④指定路由器（DR）：在一个广播性的、多接入的网络（例如 Ethernet、TokenRing 及 FDDI 环境）中，存在一个指定路由器（Designated Router，DR），指定路由器主要在 OSPF 协议中完成如下工作：

指定路由器产生用于描述所处的网段的链路数据包——Network Link，该数据包里包含该网段上所有的路由器，包括指定路由器本身的状态信息。

指定路由器与所有与其处于同一网段上的 OSPF 路由器建立相邻关系。由于 OSPF 路由器之间通过建立相邻关系及以后的 Flooding 来进行链路状态数据库是同步的，因此，我们可以说指定路由器处于一个网段的中心地位。

需要说明的是，指定路由器 DR 的定义与前面所定义的几种路由器是不同的。DR 的选择是通过 OSPF 的 Hello 数据包来完成的，在 OSPF 路由协议初始化的过程中，会通过 Hello 数据包在一个广播性网段上选出一个 ID 最大的路由器作为指定路由器 DR，并且选出 ID 次大的路由器作为备份指定路由器 BDR，BDR 在 DR 发生故障后能自动替代 DR 的所有工作。当一个网段上的 DR 和 BDR 选择好后，该网段上的其余所有路由器都只与 DR 及 BDR 建立相邻关系。在这里，一个路由器的 ID 是指向该路由器的标识，一般是有 router-id 命令指定或者是指该路由器的环回端口或是该路由器上的最大 IP 地址。

2）OSPF 链路状态广播数据包分类。

随着 OSPF 路由器种类概念的引入，OSPF 路由协议又对其链路状态广播数据包（LSA）作出了分类。OSPF 将链路状态广播数据包共分成 5 类，分别为：

类型 1：又被称为路由器链路信息数据包（Router Link），所有的 OSPF 路由器都会产生这种数据包，用于描述路由器上连接到某一个区域的链路或是某一端口的状态信息。路由器链路信息数据包只会在某一个特定的区域内广播，而不会广播至其他的区域。

在类型 1 的链路数据包中，OSPF 路由器通过对数据包中某些特定数据位的设定，告诉其余的路由器自身是一个区域边界路由器或是一个 AS 边界路由器。并且，类型 1 的链路状态数据包在描述其所连接的链路时，会根据各链路所连接的网络类型对各链路打上链路标识（Link ID）。表 17-1 列出了常见的链路类型及链路标识。

类型 2：又被称为网络链路信息数据包（Network Link）。网络链路信息数据包是由指定路由器产生的，在一个广播性的、多点接入的网络（例如以太网、令牌环网及 FDDI 网络环境）中，这种链路状态数据包用来描述该网段上所连接的所有路由器的状态信息。

表 17-1 链路类型及链路标识

| 链路类型 | 具体描述 | 链路标识 |
| --- | --- | --- |
| 1 | 用于描述点对点的网络 | 相邻路由器的路由器标识 |
| 2 | 用于描述至一个广播性网络的链路 | DR 的端口地址 |
| 3 | 用于描述至非穿透网络，即 stub 网络的链路 | stub 网络的网络号码 |
| 4 | 用于描述虚拟链路 | 相邻路由器的路由器标识 |

指定路由器 DR 只有在与至少一个路由器建立相邻关系后才会产生网络链路信息数据包，在该数据包中含有对所有已经与 DR 建立相邻关系的路由器的描述，包括 DR 路由器本身。类型 2 的链路信息只会在包含 DR 所处的广播性网络的区域中广播，不会广播至其余的 OSPF 路由区域。

类型 3 和类型 4：类型 3 和类型 4 的链路状态广播在 OSPF 路由协议中又称为总结链路信息数据包（Summary Link），该链路状态广播是由区域边界路由器或 AS 边界路由器产生的。Summary Link 描述的是到某一个区域外部的路由信息，这一个目的地址必须是在同一个 AS 中。Summary Link 也只会在某一个特定的区域内广播。类型 3 与类型 4 两种总结性链路信息的区别在于，类型 3 是由区域边界路由器产生的，用于描述到同一个 AS 中不同区域之间的链路状态；而类型 4 是由 AS 边界路由器产生的，用于描述不同 AS 的链路状态信息。

值得一提的是，只有类型 3 的 Summary Link 才能广播进一个残域，因为在一个残域中不允许存在 AS 边界路由器。残域的区域边界路由器产生一条默认的 Summary Link 对域内广播，从而在其余路由器上产生一条默认路由信息。采用 Summary Link 可以减小残域中路由器的链路状态数据库的大小，进而减少对路由器资源的利用，提高路由器的运算速度。

类型 5：又被称为 AS 外部链路状态信息数据包。类型 5 的链路数据包是由 AS 边界路由器产生的，用于描述到 AS 外的目的地的路由信息，该数据包会在 AS 中除残域以外的所有区域中广播。一般来说，这种链路状态信息描述的是到 AS 外部某一特定网络的路由信息，在这种情况下，类型 5 的链路状态数据包的链路标识采用的是目的地网络的 IP 地址；在某些情况下，AS 边界路由器可以对 AS 内部广播默认路由信息，在这时，类型 5 的链路广播数据包的链路标识采用的是默认网络号码 0.0.0.0。

（6）OSPF 协议工作过程。

OSPF 路由协议针对每一个区域分别运行一套独立的计算法则，对于区域边界路由器（ABR）来说，由于其同时与几个区域相连，因此一个区域边界路由器上会同时运行几套 OSPF 计算方法，每一个方法针对一个 OSPF 区域。下面对 OSPF 协议运算的全过程作一概括性的描述。

1）区域内部路由。

当一个 OSPF 路由器初始化时，首先初始化路由器自身的协议数据库，然后等待低层次协议（数据链路层）提示端口是否处于工作状态。

如果低层协议得知一个端口处于工作状态，OSPF 会通过其 Hello 协议数据包与其余的 OSPF 路由器建立交互关系。一个 OSPF 路由器向其相邻路由器发送 Hello 数据包时，如果接收到某一路由器返回的 Hello 数据包，则在这两个 OSPF 路由器之间建立起 OSPF 交互关系，这个过程在 OSPF 中被称为 adjacency。在广播性网络或是在点对点的网络环境中，OSPF 协议通过 Hello 数据包自动地发现其相邻路由器，此时，OSPF 路由器将 Hello 数据包发送至一特殊的多点广播地址，该多点广播地址为 ALLSPFRouters。在一些非广播性的网络环境中，我们需要经过某些设置来发现 OSPF 相邻路由器。在多接入的环境中（例如以太网的环境），Hello 协议数据包还可以用于选择该网络中的指定路由器 DR。

一个 OSPF 路由器会与其新发现的相邻路由器建立 OSPF 的 adjacency，并且在一对 OSPF 路由器之间作链路状态数据库的同步。在多接入的网络环境中，非 DR 的 OSPF 路由器只会与指定路由器 DR 建立 adjacency，并且作数据库的同步。OSPF 协议数据包的接收及发送正是在一对 OSPF 的 adjacency 间进行的。

OSPF 路由器周期性地产生与其相联的所有链路的状态信息，有时这些信息也被称为链路状态广播 LSA（Link State Advertisement）。当路由器相连接的链路状态发生改变时，路由器也会产生链路状态广播信息，所有这些广播数据是通过 Flood 的方式在某一个 OSPF 区域内进行的。Flooding 算法是一个非常可靠的计算过程，它保证在同一个 OSPF 区域内的所有路由器都具有一个相同的 OSPF 数据库。根据这个数据库，OSPF 路由器会将自身作为根，计算出一个最短路径树，然后，该路由器会根据最短路径树产生自己的 OSPF 路由表。

2）建立 OSPF 交互关系。

OSPF 路由协议通过建立交互关系来交换路由信息，但是并不是所有相邻的路由器都会建立 OSPF 交互关系。下面将 OSPF 建立 adjacency 的过程简要介绍一下。

OSPF 协议是通过 Hello 协议数据包来建立及维护相邻关系的，同时也用其来保证相邻路由器之间的双向通信。OSPF 路由器会周期性地发送 Hello 数据包，当这个路由器看到自身被列于其他路由器的 Hello 数据包里时，这两个路由器之间会建立起双向通信。在多接入的环境中，Hello 数据包还用于发现指定路由器 DR，通过 DR 来控制与哪些路由器建立交互关系。

两个 OSPF 路由器建立双向通信之后的第二个步骤是进行数据库的同步，数据库同步是所有链路状态路由协议的最大共性。在 OSPF 路由协议中，数据库同步关系仅仅在建立交互关系的路由器之间保持。

OSPF 的数据库同步是通过 OSPF 数据库描述数据包（Database Description Packets）来进行的。OSPF 路由器周期性地产生数据库描述数据包，该数据包是有序的，即附带有序列号，并将这些数据包对相邻路由器广播。相邻路由器可以根据数据库描述数据包的序列号与自身数据库的数据作比较，若发现接收到的数据比数据库内的数据序列号大，则相邻路由器会针对序列号较大的数据发出请求，并用请求得到的数据来更新其链路状态数据库。

我们可以将 OSPF 相邻路由器从发送 Hello 数据包、建立数据库同步至建立完全的 OSPF 交互关系的过程分成以下几个不同的状态：

Down：这是 OSPF 建立交互关系的初始化状态，表示在一定时间之内没有接收到从某一相邻路由器发送来的信息。在非广播性的网络环境内，OSPF 路由器还可能对处于 Down 状态的路由器发送 Hello 数据包。

Attempt：该状态仅在 NBMA 环境（例如帧中继、X.25 或 ATM 环境）中有效，表示在一定时间内没有接收到某一相邻路由器的信息，但是 OSPF 路由器仍必须通过以一个较低的频率向该相邻路由器发送 Hello 数据包来保持联系。

Init：在该状态时，OSPF 路由器已经接收到相邻路由器发送来的 Hello 数据包，但自身的 IP 地址并没有出现在该 Hello 数据包内，也就是说，双方的双向通信还没有建立起来。

2-Way：这个状态可以说是建立交互方式真正的开始步骤。在这个状态，路由器看到自身已经处于相邻路由器的 Hello 数据包内，双向通信已经建立。指定路由器及备份指定路由器的选择正是在这个状态完成的。在这个状态，OSPF 路由器还可以根据其中的一个路由器是否为指定路由器或是根据链路是否为点对点或虚拟链路来决定是否建立交互关系。

**Exstart**：这个状态是建立交互状态的第一个步骤。在这个状态，路由器要决定用于数据交换的初始的数据库描述数据包的序列号，以保证路由器得到的永远是最新的链路状态信息。同时，在这个状态，路由器还必须决定路由器之间的主备关系，处于主控地位的路由器会向处于备份地位的路由器请求链路状态信息。

**Exchange**：在这个状态，路由器向相邻的 OSPF 路由器发送数据库描述数据包来交换链路状态信息，每一个数据包都有一个数据包序列号。在这个状态，路由器还有可能向相邻路由器发送链路状态请求数据包来请求其相应数据。从这个状态开始，我们说 OSPF 处于 Flood 状态。

**Loading**：在 Loading 状态，OSPF 路由器会就其发现的相邻路由器的新链路状态数据及自身的已经过期的数据向相邻路由器提出请求，并等待相邻路由器的回答。

**Full adjacency**：这是两个 OSPF 路由器建立交互关系的最后一个状态，在这时，建立起交互关系的路由器之间已经完成了数据库同步的工作，它们的链路状态数据库已经一致。

整个建立交互关系的全过程可以用图 17-5 来表示。

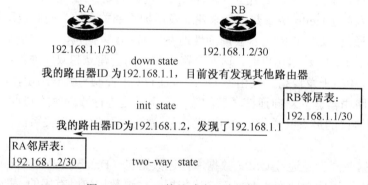

图 17-4　OSPF 协议建立双向通信过程

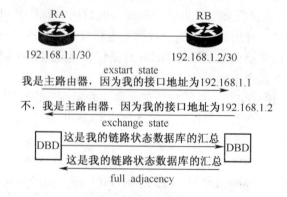

图 17-5　OSPF 协议的 adjacency 过程

3）域间路由。

在单个 OSPF 区域中，OSPF 路由协议不会产生更多的路由信息。为了与其余区域中的 OSPF 路由器通信，该区域的边界路由器会产生一些其他的信息对域内广播，这些附加信息描绘了在同一个 AS 中的其他区域的路由信息。具体路由信息交换过程如下：

在 OSPF 的定义中，所有的区域都必须与区域 0 相连，因此每一个区域都必须有一个区域边界路由器与区域 0 相连，这一个区域边界路由器会将其相连接的区域内部结构数据通过 Summary Link 广播至区域 0，也就是广播至所有其他区域的边界路由器。在这时，与区域 0 相联的边界路由器上有区域 0 及其他所有区域的链路状态信息，通过这些信息，这些边界路由器能够计算出至相应目的地的路由，并将这些路由信息广播至与其相连接的区域，以便让该区域内部的路由器找到与区域外部通信的最佳路由。

4）AS 外部路由。

一个自治域 AS 的边界路由器会将 AS 外部路由信息广播至整个 AS 中除了残域的所有区域。为了使这些 AS 外部路由信息生效，AS 内部的所有路由器（除残域内的路由器）都必须知道 AS 边界路由器的位置，该路由信息是由非残域的区域边界路由器对域内广播的，其链路广播数据包的类型为类型 4。

（7）OSPF 路由协议验证。

在 OSPF 路由协议中，所有的路由信息交换都必须经过验证。在 OSPF 协议数据包结构中，包含有一个验证域及一个 64 位长度的验证数据域，用于特定的验证方式的计算。

OSPF 数据交换的验证是基于每一个区域来定义的，也就是说，当在某一个区域的一个路由器上定义了一种验证方式时，必须在该区域的所有路由器上定义相同的协议验证方式。另外一些与验证相关的参数也可以基于每一个端口来定义，例如当采用单一口令验证时，我们可以对某一区域内部的每一个网络设置不同的口令字。

在 OSPF 路由协议的定义中，初始定义了两种协议验证方式——方式 0 及方式 1，分别介绍如下：

验证方式 0：采用验证方式 0 表示 OSPF 对所交换的路由信息不验证。在 OSPF 的数据包头内 64 位的验证数据位可以包含任何数据，OSPF 接收到路由数据后，对数据包头内的验证数据位不作任何处理。

验证方式 1：验证方式 1 为简单口令字验证。这种验证方式是基于一个区域内的每一个网络来定义的，每一个发送至该网络的数据包的包头内都必须具有相同的 64 位长度的验证数据位，也就是说验证方式 1 的口令字长度为 64bits，或者为 8 个字符。

（8）LSA 的类型。

1）类型 1（Router LSA）。

每台路由器都创建 1 类 LSA，用于向它连接的每个区域描述自己。在每台路由器中，每个区域的 LSDB 都包含一个 1 类的 LSA，它指出了当前路由器的 RID 和所有接口的 IP 地址，1 类 LSA 还用于描述末梢网路。

1 类 LSA 使用 OSPF 路由器 ID 标示 OSPF 路由器。每台路由器都创建一个 1 类的 LSA 并泛洪到整个区域。为了泛洪 LSA，始发路由器将 1 类 LSA 发送给当前区域内的邻居，然后邻居再将其发送给当前区域的其他邻居，依此类推，直到区域内的所有路由器都有该 LSA 的拷贝。

1 类 LSA 包含的信息：对于没有选举 DR 的每个接口，指出接口的子网号/掩码和 OSPF 开销；对于选举了 DR 的每个接口，指出 DR 的 IP 地址以及连接到中转网络的链路。对于没有选举 DR 但是通过它可以到达一个邻居的接口，指出该邻居的 RID。每台内部路由器都创建一个 1 类 LSA，但是 ABR 创建多个 1 类 LSA，每个区域都有一个。

2）类型 2（Network LSA）。

每个多路访问网络中，子网中的 DR 都会创建 Network LSA，描述了子网及连接到该子网的路由器接口。它只在产生这条 Network LSA 的区域泛洪描述了所有和它相连的路由器（包括 DR 本身）。

3）类型 3（Network Summary LSA）。

由 ABR 创建，描述了一个区域的 1 类和 2 类 LSA 中包含的子网，被通告到另一个区域。它指出了始发区域的链路（子网）和开销，但是没有拓扑数据。如果 ABR 知道有多条路径可以到达目标地址，但是它仍然只发送单个的 Network Summary LSA，并且是开销最低的那条；同样，如果 ABR 从其他的 ABR 那里收到多条 Network Summary LSA 的话，它会只选择开销最低的，并把这条 Network Summary LSA 宣告给其他区域。当其他的路由器收到来自 ABR 的 NetworkSummary LSA 以后，它不会运行 SPF 算法，它只简单地加上到达那个 ABR 的开销和 Network Summary LSA 中包含的开销，通过 ABR 到达目标地址的路由和开销一起被加进路由表里，这种依赖中间路由器来确定到达目标地址的完全路由（Full Route）实际上是距离矢量路由协议的行为。

4）类型 4（ASBR Summary LSA）。

类似于 3 类 LSA，只是通告一条用于前往 ASBR 的主机路由，而不是一个网络。

5）类型 5（AS External LSA）。

AS 外部 LSA，由 ASBR 创建，用于描述被注入到 OSPF 中的外部路由。这种 LSA 将在全 AS 内泛洪。

6）类型 6（Group Membership LSA）。

组成员关系 LSA，这是为 MOSPF 定义的，思科的 IoS 不支持。

7）类型 7（NSSA External LSA）。

NSSA 外部 LSA，来自非完全 Stub 区域（not-so-stubby area）内，类似于 5 类 LSA，只不过是由 NSSA 区域中的 ASBR 创建，只在 NSSA 区域内泛洪。

8）类型 8（External Attributes LSA）。

外部属性 LSA，思科路由器不能实现。

9）类型 9～类型 11（Opaque LSA）。

不透明 LSA，用作通用 LSA，以方便扩展 OSPF（如：为了支持 MPLS 流量工程而修改了类型 10 的 LSA）。

### 四、实验需要掌握的命令

router(config)#interface loopback 0
router(config-if)#ip address *ip-address subnet-mask*
#配置环回接口，设置 OSPF 路由器 ID
router(config)#router ospf *process-id*
#启动 OSPF 进程
router(config-router)#network *network-number wildcard-mask* area *area-id*
# 激活 OSPF，把网络和区域关联起来，其中 *network-number* 为网络号，*wildcard-mask* 为网络的子网掩码的反码，*area-id* 为区域号
router(config-router)#area *area-id* stub [no-summary]

#配置桩区域，*area-id* 为区域号，no-summary 选项是创建一个完全桩区域，该区域能防止把任何外部或者内部的路由引入到已经配置好的区域中

router(config-router)#area *area-id* default cost
#设置在桩区域内生成的默认路由的代价，*area-id* 为区域号，*cost* 为路由的代价

router(config-router)#area *area-id* virtual-link *router-id* [hello-interval *seconds*] [retransmit-interval *seconds*] [transmit-delay *seconds*] [dead-interval *seconds*] [[authentication] [authentication-key *key*] [message-digest-key *key-id* md5 *key*]]
#配置虚链路，提供普通区域与主干区域连通性，*area-id* 为区域号，*router-id* 为路由器 ID，hello-interval *seconds* 为设置 Hello 包的时间间隔，retransmit-interval *seconds* 为设置重传的时间间隔，transmit-delay *seconds* 为设置传输延迟的时间间隔，dead-interval *seconds* 为设置相邻设备死亡时间间隔，authentication-key *key* 为设置明文验证的口令，message-digest-key *key-id* md5 *key* 为设置加密验证的密钥

router(config-router)#area *area-id* range *summary-address mask*
#配置汇总区域间路由，*area-id* 为区域号，*summary-address* 为汇总的网段地址，*mask* 为子网掩码

router(config-if)#ip ospf priority *number*
#设置 OSPF 接口优先级，*number* 的范围是 0 到 255，默认值为 1

router(config-if)#ip ospf hello-interval *seconds*
#设置邻居 Hello 时间间隔

router(config-if)#ip ospf dead-interval *seconds*
#设置邻居死亡时间间隔

router(config-if)#ip ospf retransmit-interval *seconds*
#设置重传时间间隔

router(config-if)#ip ospf transmit-delay *seconds*
#设置传输延迟时间

router(config-if)#ip ospf cost
#设置接口代价

router(config-if)#ip ospf network {broadcast | non-broadcast |point-to-multipoint [non-broadcast]|point-to-point}
#设置 OSPF 网络的类型

router(config-router)#neighbor *ip-address* {[priority *number* [pool-interval *seconds*]] | [cost *number*]}
#设置 OSPF 的邻居，*ip-address* 为邻居的 IP 地址，priority *number* 为设置优先级，pool-interval *seconds* 为轮询时间，cost *number* 为设置邻居的代价

router(config-router)#area *area-id* authentication [message-digest]
#启用区域验证，message-digest 为 MD5 加密的验证类型

router(config-if)#ip ospf authentication
#在接口上启用明文验证

router(config-if)#ip ospf authentication-key *key*
#在接口上设置明文验证口令

router(config-if)#ip ospf authentication message-digest
#在接口上启用 MD5 加密验证

router(config-if)#ip ospf message-digest-key *keyid* md5 *key*
#在接口上设置 MD5 加密验证口令，*keyid* 的范围为 1 到 255，有相同的 *keyid* 的邻居必须有相同的 *key*

router(config-router)#redistribute *protocol* [*as-number*| *process-id*] [subnet]
#启用路由重分发，*protocol* 为动态路由选择协议，取值为 bgp、rip、eigrp、ospf、isis，*as-number* 为自治系统号，*process-id* 为进程号，subnets 为重新分发子网路由到一个 OSPF 网络时要记住的特殊信息点

router(config-router)#summary-address *ip-address mask*
#设置在外部路由重分发到 OSPF 时进行路由汇总，*ip-address* 为汇总的 IP 地址，*mask* 为子网掩码

### 五、实验拓扑图和网络文件

（1）网络拓扑图。

网络拓扑图如图 17-6 所示。

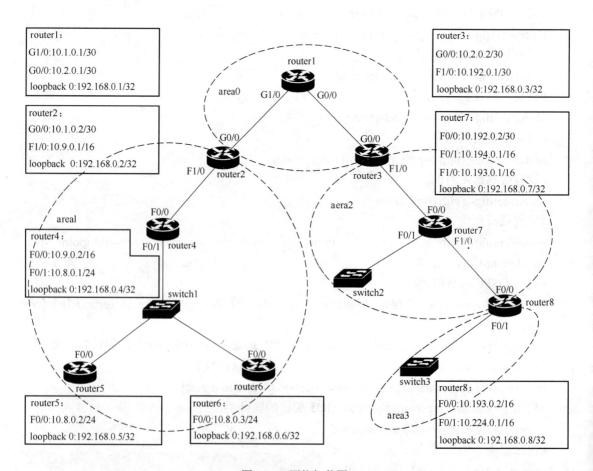

图 17-6　网络拓扑图

（2）网络文件。
# OSPF.net 文件开始
autostart = false
ghostios = true
sparsemem = true

[192.168.7.158:7200]    #本机网卡地址和端口号

    [[7200]]
    image = ..\C7200-AD.BIN
    ram = 256

    [[ROUTER R1]]
    model = 7200
    idlepc = 0xffffffff8000aed4
    slot0 = PA-C7200-IO-GE-E
    slot1 = PA-GE
    G1/0 = R2 G0/0
    G0/0 = R3 G0/0

    [[router R2]]
    model = 7200
    idlepc = 0x607c0224
    slot0 = PA-C7200-IO-GE-E
    slot1 = PA-FE-TX

    [[router R3]]
    model = 7200
    idlepc = 0xffffffff8000aed4
    slot0 = PA-C7200-IO-GE-E
    slot1 = PA-FE-TX
    F1/0 = R7 F0/0

    [[router R4]]
    model = 7200
    idlepc = 0xffffffff8000b358
    slot0 = PA-C7200-IO-2FE
    F0/0 = R2 F1/0
    F0/1 = SW1 1

[192.168.7.157:7200]

    workingdir = E:\OSPF    #在 IP 地址为 192.168.7.157 的主机上的 E 盘根目录下创建 OSPF 目录

    [[7200]]
    image = c:\program files\dynamips\images\c7200-js-mz.123-5.bin
    ram = 256

    [[router R5]]
    model = 7200
    idlepc = 0xffffffff8000aed4
    slot0 = PA-C7200-IO-2FE
    F0/0 = SW1 2


    [[router R6]]
    model = 7200
    idlepc = 0xffffffff8000aed4
    slot0 = PA-C7200-IO-2FE
    F0/0 = SW1 3


    [[router R7]]
    model = 7200
    idlepc = 0xffffffff8000b358
    slot0 = PA-C7200-IO-2FE
    slot1 = PA-FE-TX
    F1/0 = R8 F0/0
    F0/1 = SW2 1


    [[router R8]]
    model = 7200
    idlepc = 0xffffffff8000aed4
    slot0 = PA-C7200-IO-2FE
    f0/1 = SW3 1

    [[ETHSW SW1]]
    1 = access 1
    2 = access 1
    3 = access 1

    [[ETHSW SW2]]
    1 = access 1

    [[ETHSW SW3]]
    1 = access 1
# OSPF.net 文件结束

### 六、实验配置步骤

（1）路由器的基本配置步骤。
1）路由器 router1 的配置步骤。
router>enable
router#configure terminal
router(config)#hostname router1
router1(config)#interface Loopback0

router1(config-if)#ip address 192.168.0.1 255.255.255.255
router1(config-if)#interface GigabitEthernet0/0
router1(config-if)#ip address 10.2.0.1 255.255.255.252
router1(config-if)#duplex full
router1(config-if)#speed 1000
router1(config-if)#no shutdown
router1(config-if)#interface GigabitEthernet1/0
router1(config-if)#ip address 10.1.0.1 255.255.255.252
router1(config-if)#no shutdown
router1(config-if)#do write

2）路由器 router2 的配置步骤。

router>enable
router#configure terminal
router(config)#hostname router2
router2(config)#interface Loopback0
router2(config-if)#ip address 192.168.0.2 255.255.255.255
router2(config-if)#interface GigabitEthernet0/0
router2(config-if)#ip address 10.1.0.2 255.255.255.252
router2(config-if)#duplex full
router2(config-if)#speed 1000
router2(config-if)#no shutdown
router2(config-if)#interface FastEthernet1/0
router2(config-if)#ip address 10.9.0.1 255.255.0.0
router2(config-if)#duplex full
router2(config-if)#no shutdown
router2(config-if)#do write

3）路由器 router3 的配置步骤。

router>enable
router#configure terminal
router(config)#hostname router3
router3(config)#interface Loopback0
router3(config-if)#ip address 192.168.0.3 255.255.255.255
router3(config-if)#interface GigabitEthernet0/0
router3(config-if)#ip address 10.2.0.2 255.255.255.252
router3(config-if)#duplex full
router3(config-if)#speed 1000
router3(config-if)#no shutdown
router3(config-if)#interface FastEthernet1/0
router3(config-if)#ip address 10.192.0.1 255.255.255.252
router3(config-if)#duplex full
router3(config-if)#no shutdown
router3(config-if)#do write

4）路由器 router4 的配置步骤。

router>enable
router#configure terminal
router(config)#hostname router4
router4(config)#interface Loopback0

router4(config-if)#ip address 192.168.0.4 255.255.255.255
router4(config-if)#interface FastEthernet0/0
router4(config-if)#ip address 10.9.0.2 255.255.0.0
router4(config-if)#duplex full
router4(config-if)#speed 100
router4(config-if)#no shutdown
router4(config-if)#interface FastEthernet0/1
router4(config-if)#ip address 10.8.0.1 255.255.255.0
router4(config-if)#duplex full
router4(config-if)#speed 100
router4(config-if)#no shutdown
router4(config-if)#do write

5）路由器 router5 的配置步骤。

router>enable
router#configure terminal
router(config)#hostname router5
router5(config)#interface Loopback0
router5(config-if)#ip address 192.168.0.5 255.255.255.0
router5(config-if)#interface FastEthernet0/0
router5(config-if)#ip address 10.8.0.2 255.255.255.0
router5(config-if)#duplex full
router5(config-if)#speed 100
router5(config-if)#no shutdown
router5(config-if)#do write

6）路由器 router6 的配置步骤。

router>enable
router#configure terminal
router(config)#hostname router6
router6(config)#interface Loopback0
router6(config-if)#ip address 192.168.0.6 255.255.255.255
router6(config-if)#interface FastEthernet0/0
router6(config-if)#ip address 10.8.0.3 255.255.255.0
router6(config-if)#duplex full
router6(config-if)#speed 100
router6(config-if)#no shutdown
router6(config-if)#do write

7）路由器 router7 的配置步骤。

router>enable
router#configure terminal
router(config)#hostname router7
router7(config)#interface Loopback0
router7(config-if)#ip address 192.168.0.7 255.255.255.255
router7(config-if)#interface FastEthernet0/0
router7(config-if)#ip address 10.192.0.2 255.255.255.252
router7(config-if)#duplex full
router7(config-if)#speed 100
router7(config-if)#no shutdown

```
router7(config-if)#interface FastEthernet0/1
router7(config-if)#ip address 10.194.0.1 255.255.0.0
router7(config-if)#duplex full
router7(config-if)#speed 100
router7(config-if)#no shutdown
router7(config-if)#interface FastEthernet1/0
router7(config-if)#ip address 10.193.0.1 255.255.0.0
router7(config-if)#duplex full
router7(config-if)#no shutdown
router7(config-if)#do write
```

8)路由器 router8 的配置步骤。

```
router>enable
router#configure terminal
router(config)#hostname router8
router8(config)#interface Loopback0
router8(config-if)#ip address 192.168.0.8 255.255.255.255
router8(config-if)#interface FastEthernet0/0
router8(config-if)#ip address 10.193.0.2 255.255.0.0
router8(config-if)#duplex full
router8(config-if)#speed 100
router8(config-if)#no shutdown
router8(config-if)#interface FastEthernet0/1
router8(config-if)#ip address 10.224.0.1 255.255.0.0
router8(config-if)#duplex full
router8(config-if)#speed 100
router8(config-if)#no shutdown
router8(config-if)#do write
```

(2) OSPF 路由协议的配置步骤。

1)路由器 router1 的配置步骤。

```
router1(config)#router ospf 1
router1(config-router)# network 10.0.0.0 0.7.255.255 area 0
```
#此条网络宣告也可以用下面的一条网络宣告代替,但是后面生成的路由会有变化
**router1(config-router)# network 10.1.0.0 0.0..0.3area 0**
```
router1(config-router)# network 10.2.0.0 0.7.255.255 area 0
```
#此条网络宣告也可以用下面的一条网络宣告代替,但是后面生成的路由会有变化
**router1(config-router)# network 10.0.0.0 0.0..0.3area 0**
```
router1(config-router)#do write
```

2)路由器 router2 的配置步骤。

```
router2(config)#router ospf 1
router2(config-router)#area 1 range 10.8.0.0 255.248.0.0  #将子网为 10.8.0.0/13 区域间的路由进行汇总
router2(config-router)#network 10.0.0.0 0.7.255.255 area 0
```
#此条网络宣告也可以用下面的一条网络宣告代替,但是后面生成的路由会有变化
**router2(config-router)# network 10.1.0.0 0.0..0.3area 0**
```
router2(config-router)#network 10.8.0.0 0.7.255.255 area 1
```
#此条网络宣告也可以用下面的一条网络宣告代替,但是后面生成的路由会有变化
**router2(config-router)#network 10.9.0.0 0.0.255.255 area 1**
```
router2(config-router)#do write
```

3）路由器 router3 的配置步骤。

router3(config)#interface FastEthernet1/0
router3(config-if)#ip ospf authentication　#启用明文认证
router3(config-if)#ip ospf authentication-key cisco　　#设置端口认证密钥
router3(config-if)#exit
router3(config)#router ospf 1
router3(config-router)#area 2 authentication
router3(config-router)#area 2 virtual-link 192.168.0.8 authentication authentication-key cisco
#设置虚链路，同时启用虚链路认证
router3(config-router)#network 10.0.0.0 0.7.255.255 area 0
#此条网络宣告也可以用下面的一条网络宣告代替，但是后面生成的路由会有变化
**router3(config-router)#network 10.2.0.0 0.0.0.3 area 0**
router3(config-router)#network 10.192.0.0 0.7.255.255 area 2
#此条网络宣告也可以用下面的一条网络宣告代替，但是后面生成的路由会有变化
**router3(config-router)#network 10.192.0.0 0.0.0.3 area 2**
router3(config-router)#do write

4）路由器 router4 的配置步骤。

router4(config)#interface FastEthernet0/1
router4(config-if)#ip ospf priority 50
#设置端口优先级，在一个广播网络中，端口优先级最高的路由器为 DR
router4(config-if)#exit
router4(config)#router ospf 1
router4(config-router)#network 10.8.0.0 0.7.255.255 area 1
#此条网络宣告也可以用下面的两条网络宣告代替，但是后面生成的路由会有变化
**router4(config-router)#network 10.9.0.0 0.0.255.255 area 1**
**router4(config-router)#network 10.8.0.0 0.0.0.255 area 1**
router4(config-router)#do write

5）路由器 router5 的配置步骤。

router5(config)#router ospf 1
router5(config-router)#network 10.8.0.0 0.7.255.255 area 1
#此条网络宣告也可以用下面的一条网络宣告代替，但是后面生成的路由会有变化
**router5(config-router)#network 10.8.0.0 0.0.0.255 area 1**
router5(config-router)#do write

6）路由器 router6 的配置步骤。

router5(config)#interface FastEthernet0/1
router5(config-if)#ip ospf priority 0
#设置端口优先级，在一个广播网络中，端口优先级为 0 的路由器不参与 DR 的选举
router5(config-if)#exit
router6(config)#router ospf 1
router6(config-router)#network 10.8.0.0 0.7.255.255 area 1
#此条网络宣告也可以用下面的一条网络宣告代替，但是后面生成的路由会有变化
**router6(config-router)#network 10.8.0.0 0.0.0.255 area 1**
router6(config-router)#do write

7）路由器 router7 的配置步骤。

router7(config)#interface FastEthernet0/0
router7(config-if)#ip ospf authentication　　#启用明文认证
router7(config-if)#ip ospf authentication-key cisco　　#设置端口认证密钥

router7(config-if)#interface FastEthernet1/0
router7(config-if)#ip ospf authentication    #启用明文认证
router7(config-if)#ip ospf authentication-key cisco    #设置端口认证密钥
router7(config-if)#exit
router7(config)#router ospf 1
router7(config-router)#area 2 authentication    #区域启用明文认证
router7(config-router)#network 10.192.0.0 0.7.255.255 area 2
#此条网络宣告也可以用下面的三条网络宣告代替，但是后面生成的路由会有变化
**router7(config-router)#network 10.192.0.0 0.0.0.3 area 2**
**router7(config-router)#network 10.193.0.0 0.0.255.255 area 2**
**router7(config-router)#network 10.194.0.0 0.0.255.255 area 2**
router7(config-router)#do write

8）路由器 router8 的配置步骤。

router8(config)#interface FastEthernet0/0
router8(config-if)#ip ospf authentication    #启用明文认证
router8(config-if)#ip ospf authentication-key cisco    #设置端口认证密钥
router8(config-if)#exit
router8(config)#router ospf 1
router8(config-router)#area 2 authentication    #区域启用明文认证
router8(config-router)#area 2 virtual-link 192.168.0.3 authentication authentication-key cisco
#设置虚链路，同时启用虚链路认证
router8(config-router)#network 10.192.0.0 0.7.255.255 area 2
#此条网络宣告也可以用下面的一条网络宣告代替，但是后面生成的路由会有变化
**router8(config-router)#network 10.193.0.0 0.0.255.255 area 2**
router8(config-router)#network 10.224.0.0 0.7.255.255 area 3
#此条网络宣告也可以用下面的一条网络宣告代替，但是后面生成的路由会有变化
**router8(config-router)#network 10.224.0.0 0.0.255.255 area 3**
router8(config-router)#do write

## 七、实验测试和查看相关配置

（1）查看路由器 router1 的路由信息。

router1#show ip route
O IA    10.8.0.0/13 [110/2] via 10.1.0.2, 00:37:31, GigabitEthernet1/0 ：路由中 O 的意思是此条路由是由 OSPF 路由协议生成的动态路由；IA 的意思是此路由是由 OSPF 其他区域生成的路由，通过域间路由器传递过来的域间路由；10.8.0.0/13，其中 10.8.0.0 是目的网段地址，/13 的意思是目的网段子网掩码前缀为 13 位；[110/2]的意思是管理距离为 110，度量值为 2；GigabitEthernet1/0 的意思是下一跳的接口为 GigabitEthernet1/0。本条路由的意思是要到达目的网段 10.8.0.0/13 子网需要通过 GigabitEthernet1/0 接口进行转发。

Codes: C - connected, S - static, R - RIP, M - mobile, B - BGP
       D - EIGRP, EX - EIGRP external, O - OSPF, IA - OSPF inter area
       N1 - OSPF NSSA external type 1, N2 - OSPF NSSA external type 2
       E1 - OSPF external type 1, E2 - OSPF external type 2
       i - IS-IS, su - IS-IS summary, L1 - IS-IS level-1, L2 - IS-IS level-2
       ia - IS-IS inter area, * - candidate default, U - per-user static route
       o - ODR, P - periodic downloaded static route

Gateway of last resort is not set

       10.0.0.0/8 is variably subnetted, 7 subnets, 3 masks
O IA    10.8.0.0/13 [110/2] via 10.1.0.2, 00:37:31, GigabitEthernet1/0
C       10.2.0.0/30 is directly connected, GigabitEthernet0/0
C       10.1.0.0/30 is directly connected, GigabitEthernet1/0
O IA    10.194.0.0/16 [110/3] via 10.2.0.2, 00:03:58, GigabitEthernet0/0
O IA    10.192.0.0/30 [110/2] via 10.2.0.2, 00:03:53, GigabitEthernet0/0
O IA    10.193.0.0/16 [110/3] via 10.2.0.2, 00:03:58, GigabitEthernet0/0
O IA    10.224.0.0/16 [110/4] via 10.2.0.2, 00:37:31, GigabitEthernet0/0
       192.168.0.0/32 is subnetted, 1 subnets
C       192.168.0.1 is directly connected, Loopback0

（2）查看路由器 router2 的路由信息。

router2# show ip route

O     10.8.0.0/13 is a summary, 00:30:37, Null0 ：路由中 O 的意思是此条路由是由 OSPF 路由协议生成的动态路由； 10.8.0.0/13，其中 10.8.0.0 是目的网段地址，/13 的意思是目的网段子网掩码前缀为 13 位；is a summary 代表此条路由是一条汇总路由；Null0 的意思是下一跳的接口为 Null0，Null0 接口是一个空接口。本条路由的意思是要到达目的网段 10.8.0.0/13 子网需要通过 Null0 接口进行转发，即到达目的网段 10.8.0.0/13 子网的数据包在此路由器上进行丢弃。此例中到达目的网段 10.8.0.0/13 子网的数据包并不是全部丢弃，而是到达 10.8.0.0/24 子网的数据包通过"O     10.8.0.0/24 [110/2] via 10.9.0.2, 00:30:37, FastEthernet1/0"路由进行转发。这和路由转发的规则有关，详细内容见静态路由章节。

Codes: C - connected, S - static, R - RIP, M - mobile, B - BGP
       D - EIGRP, EX - EIGRP external, O - OSPF, IA - OSPF inter area
       N1 - OSPF NSSA external type 1, N2 - OSPF NSSA external type 2
       E1 - OSPF external type 1, E2 - OSPF external type 2
       i - IS-IS, su - IS-IS summary, L1 - IS-IS level-1, L2 - IS-IS level-2
       ia - IS-IS inter area, * - candidate default, U - per-user static route
       o - ODR, P - periodic downloaded static route

Gateway of last resort is not set

       10.0.0.0/8 is variably subnetted, 9 subnets, 4 masks
O       10.8.0.0/24 [110/2] via 10.9.0.2, 00:30:37, FastEthernet1/0
O       10.8.0.0/13 is a summary, 00:30:37, Null0
C       10.9.0.0/16 is directly connected, FastEthernet1/0
O       10.2.0.0/30 [110/2] via 10.1.0.1, 00:38:44, GigabitEthernet0/0
C       10.1.0.0/30 is directly connected, GigabitEthernet0/0
O IA    10.194.0.0/16 [110/4] via 10.1.0.1, 00:05:11, GigabitEthernet0/0
O IA    10.192.0.0/30 [110/3] via 10.1.0.1, 00:05:06, GigabitEthernet0/0
O IA    10.193.0.0/16 [110/4] via 10.1.0.1, 00:05:11, GigabitEthernet0/0
O IA    10.224.0.0/16 [110/5] via 10.1.0.1, 00:30:37, GigabitEthernet0/0
       192.168.0.0/32 is subnetted, 1 subnets
C       192.168.0.2 is directly connected, Loopback0

（3）查看路由器 router3 的路由信息。

router3#show ip route

Codes: C - connected, S - static, R - RIP, M - mobile, B - BGP
       D - EIGRP, EX - EIGRP external, O - OSPF, IA - OSPF inter area
       N1 - OSPF NSSA external type 1, N2 - OSPF NSSA external type 2

　　　　E1 - OSPF external type 1, E2 - OSPF external type 2
　　　　i - IS-IS, su - IS-IS summary, L1 - IS-IS level-1, L2 - IS-IS level-2
　　　　ia - IS-IS inter area, * - candidate default, U - per-user static route
　　　　o - ODR, P - periodic downloaded static route

Gateway of last resort is not set

　　　　10.0.0.0/8 is variably subnetted, 7 subnets, 3 masks
O IA　　10.8.0.0/13 [110/3] via 10.2.0.1, 00:05:57, GigabitEthernet0/0
C　　　 10.2.0.0/30 is directly connected, GigabitEthernet0/0
O　　　 10.1.0.0/30 [110/2] via 10.2.0.1, 00:05:57, GigabitEthernet0/0
O　　　 10.194.0.0/16 [110/2] via 10.192.0.2, 00:05:57, FastEthernet1/0
C　　　 10.192.0.0/30 is directly connected, FastEthernet1/0
O　　　 10.193.0.0/16 [110/2] via 10.192.0.2, 00:05:57, FastEthernet1/0
O IA　　10.224.0.0/16 [110/3] via 10.192.0.2, 00:05:57, FastEthernet1/0
　　　　192.168.0.0/32 is subnetted, 1 subnets
C　　　 192.168.0.3 is directly connected, Loopback0

（4）查看路由器 router4 的路由信息。
router4#show ip route
Codes: C - connected, S - static, R - RIP, M - mobile, B - BGP
　　　　D - EIGRP, EX - EIGRP external, O - OSPF, IA - OSPF inter area
　　　　N1 - OSPF NSSA external type 1, N2 - OSPF NSSA external type 2
　　　　E1 - OSPF external type 1, E2 - OSPF external type 2
　　　　i - IS-IS, su - IS-IS summary, L1 - IS-IS level-1, L2 - IS-IS level-2
　　　　ia - IS-IS inter area, * - candidate default, U - per-user static route
　　　　o - ODR, P - periodic downloaded static route

Gateway of last resort is not set

　　　　10.0.0.0/8 is variably subnetted, 8 subnets, 3 masks
C　　　 10.8.0.0/24 is directly connected, FastEthernet0/1
C　　　 10.9.0.0/16 is directly connected, FastEthernet0/0
O IA　　10.2.0.0/30 [110/3] via 10.9.0.1, 00:32:02, FastEthernet0/0
O IA　　10.1.0.0/30 [110/2] via 10.9.0.1, 00:32:02, FastEthernet0/0
O IA　　10.194.0.0/16 [110/5] via 10.9.0.1, 00:32:02, FastEthernet0/0
O IA　　10.192.0.0/30 [110/4] via 10.9.0.1, 00:06:31, FastEthernet0/0
O IA　　10.193.0.0/16 [110/5] via 10.9.0.1, 00:32:02, FastEthernet0/0
O IA　　10.224.0.0/16 [110/6] via 10.9.0.1, 00:32:02, FastEthernet0/0
　　　　192.168.0.0/32 is subnetted, 1 subnets
C　　　 192.168.0.4 is directly connected, Loopback0

（5）查看路由器 router5 的路由信息。
router5#show ip route
Codes: C - connected, S - static, R - RIP, M - mobile, B - BGP
　　　　D - EIGRP, EX - EIGRP external, O - OSPF, IA - OSPF inter area
　　　　N1 - OSPF NSSA external type 1, N2 - OSPF NSSA external type 2
　　　　E1 - OSPF external type 1, E2 - OSPF external type 2
　　　　i - IS-IS, su - IS-IS summary, L1 - IS-IS level-1, L2 - IS-IS level-2

```
         ia - IS-IS inter area, * - candidate default, U - per-user static route
         o - ODR, P - periodic downloaded static route

Gateway of last resort is not set

         10.0.0.0/8 is variably subnetted, 8 subnets, 3 masks
C        10.8.0.0/24 is directly connected, FastEthernet0/0
O        10.9.0.0/16 [110/2] via 10.8.0.1, 00:13:04, FastEthernet0/0
O IA     10.2.0.0/30 [110/4] via 10.8.0.1, 00:13:04, FastEthernet0/0
O IA     10.1.0.0/30 [110/3] via 10.8.0.1, 00:13:04, FastEthernet0/0
O IA     10.194.0.0/16 [110/6] via 10.8.0.1, 00:13:04, FastEthernet0/0
O IA     10.192.0.0/30 [110/5] via 10.8.0.1, 00:02:53, FastEthernet0/0
O IA     10.193.0.0/16 [110/6] via 10.8.0.1, 00:13:04, FastEthernet0/0
O IA     10.224.0.0/16 [110/7] via 10.8.0.1, 00:13:04, FastEthernet0/0
C        192.168.0.0/24 is directly connected, Loopback0
```

（6）查看路由器 router6 的路由信息。

```
router6#show ip route
Codes: C - connected, S - static, R - RIP, M - mobile, B - BGP
       D - EIGRP, EX - EIGRP external, O - OSPF, IA - OSPF inter area
       N1 - OSPF NSSA external type 1, N2 - OSPF NSSA external type 2
       E1 - OSPF external type 1, E2 - OSPF external type 2
       i - IS-IS, su - IS-IS summary, L1 - IS-IS level-1, L2 - IS-IS level-2
       ia - IS-IS inter area, * - candidate default, U - per-user static route
       o - ODR, P - periodic downloaded static route

Gateway of last resort is not set

         10.0.0.0/8 is variably subnetted, 8 subnets, 3 masks
C        10.8.0.0/24 is directly connected, FastEthernet0/0
O        10.9.0.0/16 [110/2] via 10.8.0.1, 00:26:55, FastEthernet0/0
O IA     10.2.0.0/30 [110/4] via 10.8.0.1, 00:26:55, FastEthernet0/0
O IA     10.1.0.0/30 [110/3] via 10.8.0.1, 00:26:55, FastEthernet0/0
O IA     10.194.0.0/16 [110/6] via 10.8.0.1, 00:26:55, FastEthernet0/0
O IA     10.192.0.0/30 [110/5] via 10.8.0.1, 00:06:22, FastEthernet0/0
O IA     10.193.0.0/16 [110/6] via 10.8.0.1, 00:26:55, FastEthernet0/0
O IA     10.224.0.0/16 [110/7] via 10.8.0.1, 00:26:55, FastEthernet0/0
         192.168.0.0/32 is subnetted, 1 subnets
C        192.168.0.6 is directly connected, Loopback0
```

（7）查看路由器 router7 的路由信息。

```
router7#show ip route
Codes: C - connected, S - static, R - RIP, M - mobile, B - BGP
       D - EIGRP, EX - EIGRP external, O - OSPF, IA - OSPF inter area
       N1 - OSPF NSSA external type 1, N2 - OSPF NSSA external type 2
       E1 - OSPF external type 1, E2 - OSPF external type 2
       i - IS-IS, su - IS-IS summary, L1 - IS-IS level-1, L2 - IS-IS level-2
       ia - IS-IS inter area, * - candidate default, U - per-user static route
       o - ODR, P - periodic downloaded static route
```

Gateway of last resort is not set

```
      10.0.0.0/8 is variably subnetted, 7 subnets, 3 masks
O IA    10.8.0.0/13 [110/4] via 10.192.0.1, 00:34:13, FastEthernet0/0
O       10.2.0.0/30 [110/2] via 10.192.0.1, 00:34:13, FastEthernet0/0
O       10.1.0.0/30 [110/3] via 10.192.0.1, 00:34:13, FastEthernet0/0
C       10.194.0.0/16 is directly connected, FastEthernet0/1
C       10.192.0.0/30 is directly connected, FastEthernet0/0
C       10.193.0.0/16 is directly connected, FastEthernet1/0
O IA    10.224.0.0/16 [110/2] via 10.193.0.2, 00:34:13, FastEthernet1/0
      192.168.0.0/32 is subnetted, 1 subnets
C       192.168.0.7 is directly connected, Loopback0
```

（8）查看路由器 router8 的路由信息。

```
router8#show ip route
Codes: C - connected, S - static, R - RIP, M - mobile, B - BGP
       D - EIGRP, EX - EIGRP external, O - OSPF, IA - OSPF inter area
       N1 - OSPF NSSA external type 1, N2 - OSPF NSSA external type 2
       E1 - OSPF external type 1, E2 - OSPF external type 2
       i - IS-IS, su - IS-IS summary, L1 - IS-IS level-1, L2 - IS-IS level-2
       ia - IS-IS inter area, * - candidate default, U - per-user static route
       o - ODR, P - periodic downloaded static route
```

Gateway of last resort is not set

```
      10.0.0.0/8 is variably subnetted, 7 subnets, 3 masks
O IA    10.8.0.0/13 [110/5] via 10.193.0.1, 00:12:00, FastEthernet0/0
O       10.2.0.0/30 [110/3] via 10.193.0.1, 00:12:00, FastEthernet0/0
O       10.1.0.0/30 [110/4] via 10.193.0.1, 00:12:00, FastEthernet0/0
O       10.194.0.0/16 [110/2] via 10.193.0.1, 00:12:00, FastEthernet0/0
O       10.192.0.0/30 [110/2] via 10.193.0.1, 00:12:00, FastEthernet0/0
C       10.193.0.0/16 is directly connected, FastEthernet0/0
C       10.224.0.0/16 is directly connected, FastEthernet0/1
      192.168.0.0/32 is subnetted, 1 subnets
C       192.168.0.8 is directly connected, Loopback0
```

（9）查看 router4、router5、router6 的邻居信息。

Neighbor ID 字段为邻居路由器的 ID 号；Pri 字段为邻居路由器相应接口的优先级；State 字段为邻居路由器的状态，FULL 为全连接状态，DR 为此网络的指定路由，BDR 为备份指定路由，DROTHER 说明此路由器既不是指定路由器，也不是备份指定路由器；Dead Time 字段为 DR 的过期时间；Address 字段为邻居路由器的接口 IP 地址；Interface 字段为邻居路由器的接口。

```
router4#show ip ospf neighbor
Neighbor ID     Pri   State            Dead Time   Address      Interface
192.168.0.5     1     FULL/BDR         00:00:06    10.8.0.2     FastEthernet0/1
192.168.0.6     0     FULL/DROTHER     00:00:17    10.8.0.3     FastEthernet0/1
192.168.0.2     1     FULL/BDR         00:00:24    10.9.0.1     FastEthernet0/0

router5#show ip ospf neighbor
```

| Neighbor ID | Pri | State | Dead Time | Address | Interface |
|---|---|---|---|---|---|
| 192.168.0.4 | 50 | FULL/DR | 00:00:28 | 10.8.0.1 | FastEthernet0/0 |
| 192.168.0.6 | 0 | FULL/DROTHER | 00:00:24 | 10.8.0.3 | FastEthernet0/0 |

router6#show ip ospf neighbor

| Neighbor ID | Pri | State | Dead Time | Address | Interface |
|---|---|---|---|---|---|
| 192.168.0.4 | 50 | FULL/DR | 00:00:23 | 10.8.0.1 | FastEthernet0/0 |
| 192.168.0.5 | 1 | FULL/BDR | 00:00:19 | 10.8.0.2 | FastEthernet0/0 |

（10）查看路由器 router3 的虚链路的配置信息。

router3#show ip ospf virtual-links

Virtual Link OSPF_VL3 to router 192.168.0.8 is up　　#到 192.168.0.8 路由器的虚链路正常

　Run as demand circuit

　DoNotAge LSA allowed.

　Transit area 2, via interface FastEthernet1/0, Cost of using 2　　#虚链路穿过 area 2 区域

　Transmit Delay is 1 sec, State POINT_TO_POINT, #虚链路通过点对点的方式进行连接

　Timer intervals configured, Hello 10, Dead 40, Wait 40, Retransmit 5

　　Hello due in 00:00:03

　　Adjacency State FULL (Hello suppressed)

　　Index 2/3, retransmission queue length 0, number of retransmission 1

　　First 0x0(0)/0x0(0) Next 0x0(0)/0x0(0)

　　Last retransmission scan length is 1, maximum is 1

　　Last retransmission scan time is 0 msec, maximum is 0 msec

　Simple password authentication enabled

router8#show ip ospf virtual-links

Virtual Link OSPF_VL4 to router 192.168.0.3 is up

　Run as demand circuit

　DoNotAge LSA allowed.

　Transit area 2, via interface FastEthernet0/0, Cost of using 2

　Transmit Delay is 1 sec, State POINT_TO_POINT,

　Timer intervals configured, Hello 10, Dead 40, Wait 40, Retransmit 5

　　Hello due in 00:00:02

　　Adjacency State FULL (Hello suppressed)

　　Index 1/2, retransmission queue length 0, number of retransmission 1

　　First 0x0(0)/0x0(0) Next 0x0(0)/0x0(0)

　　Last retransmission scan length is 1, maximum is 1

　　Last retransmission scan time is 0 msec, maximum is 0 msec

　Simple password authentication enabled

## 八、注意问题

（1）配置 OSPF 动态路由验证时，要保持验证两端的密钥链和模式一致。

（2）在配置路由汇总时，要保证动态生成的路由正确，否则要关闭路由汇总。

（3）在配置多种路由选择协议时，一定要用路由注入的命令，而且是每种路由选择协议之间要互相注入。

# 实验十八　策略路由的配置

策略路由是一种比基于目标网络进行路由更加灵活的数据包路由转发机制。应用了策略路由，路由器将通过路由图决定如何对需要路由的数据包进行处理，路由图决定了一个数据包的下一跳转发路由器。策略路由可以使数据包按照用户指定的策略进行转发。对于某些管理目的，如 QoS 需求或 VPN 拓扑结构，要求某些特定的数据包必须经过特定的路径，就可以使用策略路由。

## 一、实验目的

（1）理解策略路由的工作原理；
（2）掌握策略路由的配置。

## 二、实验内容

（1）策略路由的配置与应用；
（2）路由控制图的配置与应用。

## 三、实验原理

传统的路由策略都是使用从路由协议派生出来的路由表，根据目的地址进行报文的转发。在这种机制下，路由器只能根据报文的目的地址为用户提供比较单一的路由方式，它更多的是解决网络数据的转发问题，而不能提供有差别的服务。

基于策略的路由为网络管理者提供了比传统路由协议对报文的转发和存储更强的控制能力。基于策略的路由比传统路由控制能力更强，使用更灵活，它使网络管理者不仅能够根据目的地址，而且能够根据协议类型、报文大小、网络管理应用、IP 源地址或者其他的策略来选择转发路径。策略可以根据实际网络管理应用的需要进行定义来控制多个路由器之间的负载均衡、单一链路上报文转发的 QoS 或者满足某种特定需求。

策略路由提供了这样一种机制：根据网络管理者制定的标准来进行报文的转发。这种标准根据实际的网络管理应用需求来制定，它的依据可以是协议类型、网络管理应用、报文大小或者 IP 源地址中的一个或者多个的组合。

当数据包经过路由器转发时，路由器根据预先设定的策略对数据包进行匹配，如果匹配到一条策略，就根据该条策略指定的路由进行转发；如果没有匹配到任何策略，就使用路由表中的各项，根据目的地址对报文进行路由。

常见的网络管理应用模式有以下几种：

（1）上网负载均衡：对于多条 ISP 线路，网络管理员可以在不同的路径之间根据带宽分配内网的上网流量，实现负载平衡。

（2）基于源地址选路：例如一个网络通过两条速度不同的线路接入互联网，管理员可以指定内网中一些特定的用户使用快速线路，而普通用户使用慢速线路。

（3）根据服务级别选路：对于不同服务要求（如传送速率、吞吐量、可靠性等）的数据，根据网络的状况进行不同的路由。如指定语音与视频等网络管理应用走带宽大的线路，数据网络管理应用走带宽小的线路。

（4）VPN 线路备份：安全网关的几条接入线路都可以作为 VPN 接入的端点，移动用户可以从任何一个接入点通过 VPN 隧道安全接入内网，从而实现 VPN 接入线路的相互备份。

## 四、实验需要掌握的命令

Router(config-if)#ip policy router-map *name*
#将路由控制图应用到用于策略路由的一个接口上
router(config-route-map)#match ip address {*ACL number|name*} [···*ACL number|name*]
#匹配 ACL 所指定的数据包的特征路由
router(config-route-map)#match length {*min*} {*max*}
#匹配网络层数据包的长度
router(config-route-map)#set default interface {*type number*} [···*type number*]
#当不存在指向目标网络的显式路由（explicit route）的时候，为匹配成功的数据包设置出口接口
router(config-route-map)#set interface {*type number*} [···*type number*]
#当存在指向目标网络的显式路由的时候，为匹配成功的数据包设置出口接口
router(config-route-map)#set ip default next-hop {*ip-address*} [···*ip-address*]
#当不存在指向目标网络的显式路由的时候，为匹配成功的数据包设置下一跳路由器地址
router(config-route-map)#set ip precedence {precedence}
#为匹配成功的 IP 数据包设置服务类型（Type of Service,ToS）的优先级
router(config-route-map)#set ip tos {tos}
#为匹配成功的数据包设置服务类型的字段的 ToS 位
router#show ip policy
#查看策略路由的配置信息
router#debug ip policy
#开启策略路由调试
router#show debugging
#显示策略路由调试信息
router#show route-map *name*
#显示路由控制图信息
Router(config)#route-map *name* {permit|deny} {*ACL_number|ACL_name*}
#定义路由控制图，*name* 为路由控制图的名称，动作 permit 为允许，动作 deny 为丢弃，*ACL_number|name* 为 ACL 编号，*ACL_name* 为 ACL 名称

## 五、实验拓扑图和网络文件

（1）网络拓扑图。
网络拓扑图如图 18-1 所示。
（2）网络文件。
#PBR.net 文件开始
autostart = false

实验十八 策略路由的配置

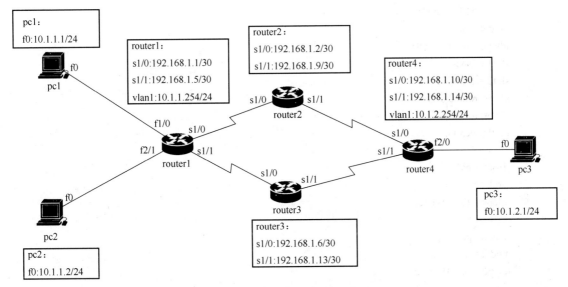

图 18-1 策略路由网络拓扑图

```
ghostios = true
sparsemem = true

[localhost]

    [[3660]]
    image = ..\C3660-IS.BIN
    ram = 160

    [[1720]]
    image = ..\C1700-Y.BIN
    ram = 16

    [[ROUTER route1]]
    model = 3660
    idlepc = 0x60719b98
    slot1 = NM-4T
    slot2 = NM-16ESW
    s1/0 = route2 s1/0
    s1/1 = route3 s1/0
    F2/0 = pc1 F0
    F2/1 = pc2 F0

    [[ROUTER route2]]
    model = 3660
    idlepc = 0x60719b98
    slot1 = NM-4T

    [[ROUTER route3]]
    model = 3660
    idlepc = 0x60719b98
    slot1 = NM-4T
```

```
[[ROUTER route4]]
model = 3660
idlepc = 0x60719b98
slot1 = NM-4T
slot2 = NM-16ESW
s1/0 = route2 s1/1
s1/1 = route3 s1/1
F2/0 = pc3 F0

[[router pc1]]
model = 1720
idlepc = 0x8014e4ec

[[router pc2]]
model = 1720
idlepc = 0x8014e4ec

[[router pc3]]
model = 1720
idlepc = 0x8014e4ec
#PBR.net 文件结束
```

## 六、实验配置步骤

（1）基本配置。

1）router1 的基本配置步骤。

```
router>enable
router#configure terminal
router(config)#hostname Router1
router1(config)#interface Serial1/0
router1(config-if)#ip address 192.168.1.1 255.255.255.252
router1(config-if)#clock rate 9600
router1(config-if)#ip nat outside
router1(config-if)#no shutdown
router1(config-if)#interface Serial1/1
router1(config-if)#ip address 192.168.1.5 255.255.255.252
router1(config-if)#clock rate 9600
router1(config-if)#ip nat outside
router1(config-if)#no shutdown
router1(config-if)#interface Vlan1
router1(config-if)#ip address 10.1.1.254 255.255.255.0
router1(config-if)#ip nat inside
router1(config-if)#no shutdown
router1(config-if)#exit
router1(config)#access-list 1 permit 10.1.1.1
router1(config)#access-list 2 permit 10.1.1.2
router1(config)#ip nat inside source list 1 interface Serial1/0 overload
router1(config)#ip nat inside source list 2 interface Serial1/1 overload
```

router1(config)#do write

2）router2 的基本配置步骤。

router>enable
router#configure terminal
router(config)#hostname Router2
router2(config)#interface Serial1/0
router2(config-if)#ip address 192.168.1.2 255.255.255.252
router2(config-if)#clock rate 9600
router2(config-if)#no shutdown
router2(config-if)#interface Serial1/1
router2(config-if)#ip address 192.168.1.9 255.255.255.252
router2(config-if)#clock rate 9600
router2(config-if)#no shutdown
router2(config-if)#exit
router2(config)#router ospf 1
router2(config-router)#network 192.168.1.0 0.0.0.15 area 0
router2(config-router)#do write

3）router3 的基本配置步骤。

router>enable
router#configure terminal
router(config)#hostname Router3
router3(config)#interface Serial1/0
router3(config-if)#ip address 192.168.1.6 255.255.255.252
router3(config-if)#clock rate 9600
router3(config-if)#no shutdown
router3(config-if)#interface Serial1/1
router3(config-if)#ip address 192.168.1.13 255.255.255.252
router3(config-if)#clock rate 9600
router3(config-if)#no shutdown
router3(config-if)#exit
router3(config)#router ospf 1
router3(config-router)#network 192.168.1.0 0.0.0.15 area 0
router3(config-router)#do write

4）router4 的基本配置步骤。

router>enable
router#configure terminal
router(config)#hostname Router4
router4(config)#interface Serial1/0
router4(config-if)#ip address 192.168.1.10 255.255.255.252
router4(config-if)#clock rate 9600
router4(config-if)#no shutdown
router4(config-if)#interface Serial1/1
router4(config-if)#ip address 192.168.1.14 255.255.255.252
router4(config-if)#clock rate 9600
router4(config-if)#no shutdown
router4(config-if)#interface Vlan1
router4(config-if)#ip address 10.1.2.254 255.255.255.0

router4(config-if)#no shutdown
router4(config-if)#exit
router4(config)#router ospf 1
router4(config-router)#network 192.168.1.0 0.0.0.15 area 0
router4(config-router)#network 10.1.2.0 0.0.0.255 area 0
router4(config-router)#do write

5) pc1 的基本配置步骤。

router>enable
router#configure terminal
router(config)#hostname pc1
pc1(config)#interface FastEthernet0
pc1(config-if)#ip address 10.1.1.1 255.255.255.0
pc1(config-if)#full-duplex
pc1(config-if)#exit
pc1(config)#no ip routing
pc1(config)#ip default-gateway 10.1.1.254
pc1(config)#do write

6) pc2 的基本配置步骤。

router>enable
router#configure terminal
router(config)#hostname pc2
pc2(config)#interface FastEthernet0
pc2(config-if)#ip address 10.1.1.2 255.255.255.0
pc2(config-if)#full-duplex
pc2(config-if)#exit
pc2(config)#no ip routing
pc2(config)#ip default-gateway 10.1.1.254
pc2(config)#do write

7) pc3 的基本配置步骤。

router>enable
router#configure terminal
router(config)#hostname pc3
pc3(config)#interface FastEthernet0
pc3(config-if)#ip address 10.1.2.1 255.255.255.0
pc3(config-if)#full-duplex
pc3(config-if)#exit
pc3(config)#no ip routing
pc3(config)#ip default-gateway 10.1.2.254
pc3(config)#do write

（2）策略路由的配置。

Router1#configure terminal
router1(config)#access-list 3 deny any
router1(config)#route-map pbr1 permit 10
router1(config-route-map)#match ip address 1
router1(config-route-map)#set ip next-hop 192.168.1.2
router1(config-route-map)#route-map pbr1 permit 15
router1(config-route-map)#match ip address 2

router1(config-route-map)#set ip next-hop 192.168.1.6
router1(config-route-map)#route-map pbr1 permit 20
router1(config-route-map)#match ip address 3
router1(config)#interface Vlan1
router1(config-if)# ip policy route-map pbr1
router1(config-if)#do write

## 七、实验测试和查看相关配置

（1）查看策略路由的配置信息。

router1#show ip policy
Interface 字段为应用策略路由的接口；Route map 字段为路由控制图
Interface          Route map
Vlan1              pbr1

（2）开启策略路由调试开关。

router1#debug ip policy
Policy routing debugging is on

（3）显示策略路由调试信息。

router1#show debugging
Policy Routing:
    Policy routing debugging is on
*Mar   1 07:34:27.618: IP: s=10.1.1.1 (Vlan1), d=10.1.2.1, len 100, FIB policy match
*Mar   1 07:34:27.622: IP: s=10.1.1.1 (Vlan1), d=10.1.2.1, g=192.168.1.2, len 100, FIB policy routed
*Mar   1 07:34:27.990: IP: s=10.1.1.1 (Vlan1), d=10.1.2.1, len 100, FIB policy match
*Mar   1 07:34:27.990: IP: s=10.1.1.1 (Vlan1), d=10.1.2.1, g=192.168.1.2, len 100, FIB policy routed
……

（4）显示 NAT 地址转换的配置信息。

router1#show ip nat statistics
Total active translations: 0 (0 static, 0 dynamic; 0 extended)
Outside interfaces:
    Serial1/0, Serial1/1
Inside interfaces:
    Vlan1
Hits: 70   Misses: 136
CEF Translated packets: 198, CEF Punted packets: 0
Expired translations: 139
Dynamic mappings:
-- Inside Source
[Id: 5] access-list 1 interface Serial1/0 refcount 0
[Id: 6] access-list 2 interface Serial1/1 refcount 0
Queued Packets: 0

（5）显示路由控制图的配置信息。

router1#show route-map pbr1
route-map pbr1, permit, sequence 10    #路由控制图 pbr1，动作为允许，编号为 10
    Match clauses:     #匹配的访问列表
      ip address (access-lists): 1
    Set clauses:    #设置的下一条地址

```
        ip next-hop 192.168.1.2
    Policy routing matches: 193 packets, 24018 bytes
route-map pbr1, permit, sequence 15    #路由控制图 pbr1，动作为允许，编号为 15
    Match clauses:
        ip address (access-lists): 2
    Set clauses:
        ip next-hop 192.168.1.6
    Policy routing matches: 121 packets, 15198 bytes
route-map pbr1, permit, sequence 20    #路由控制图 pbr1，动作为允许，编号为 20
    Match clauses:
        ip address (access-lists): 3
    Set clauses:
    Policy routing matches: 0 packets, 0 bytes
```

## 八、注意问题

（1）配置 NAT 时要注意内部端口和外部端口，哪些内部网络经过哪个端口做地址转换。

（2）在配置路由控制图时，要分清 ACL 定义的网络从路由器哪个端口进行转发，以及转发的动作是丢弃还是允许。

（3）要分清策略路由该应用到哪个端口上。

# 实验十九　PPP 协议的配置

PPP（Point-to-Point Protocol，点到点协议）是为在同等单元之间传输数据包这样的简单链路设计的链路层协议。这种链路提供全双工操作，并按照顺序传递数据包。设计目的主要是用来通过拨号或专线方式建立点对点连接发送数据，使其成为各种主机、网桥和路由器之间简单连接的一种共通的解决方案。

## 一、实验目的

理解 PPP 协议的工作原理，掌握 PPP 协议 PAP、CHAP 的配置。

## 二、实验内容

PPP 协议 PAP、CHAP 的配置。

## 三、实验原理

（1）PPP 协议的概述。

点到点协议（Point to Point Protocol，PPP）是 IETF（Internet Engineering Task Force，因特网工程任务组）推出的点到点类型线路的数据链路层协议。它解决了 SLIP 中的问题，并成为正式的因特网标准。

PPP 支持在各种物理类型的点到点串行线路上传输上层协议报文。PPP 有很多丰富的可选特性，如支持多协议、提供可选的身份认证服务、可以以各种方式压缩数据、支持动态地址协商、支持多链路捆绑等，这些丰富的选项增强了 PPP 的功能。同时，不论是异步拨号线路还是路由器之间的同步链路均可使用。

（2）PPP 协议的结构。

封装：PPP 封装提供了不同网络层协议同时通过统一链路的多路技术。精心地设计 PPP 封装，使其保有对常用支持硬件的兼容性。当使用默认的类 HDLC 帧（HDLC-like Framing）时，仅需要 8 个额外的字节就可以形成封装。在带宽需要付费时，封装和帧可以减少到 2 或 4 个字节。为了支持高速的执行，默认的封装只使用简单的字段，多路分解只需要对其中的一个字段进行检验。默认的头和信息字段落在 32bit 边界上，尾字节可以被填补到任意的边界。

链路控制协议（LCP）：为了在一个很宽广的环境内能足够方便地使用，PPP 提供了 LCP。LCP 用于就封装格式选项自动地达成一致，处理数据包大小的变化，探测 looped-back 链路和其他普通的配置错误，以及终止链路。提供的其他可选设备有：对链路中同等单元标识的认证和当链路功能正常或链路失败时的决定。

网络控制协议（NCP）：点对点连接可能和当前的一簇网络协议产生许多问题。例如，基于电路交换的点对点连接（比如拨号模式服务），分配和管理 IP 地址，即使在 LAN 环境中，也非常困难。这些问题由一簇网络控制协议（NCP）来处理，每一个协议管理着各自的网络层协议的特殊需求。

(3) PPP 帧格式。

PPP 帧格式和 HDLC 帧格式相似，如图 19-1 所示。二者主要区别：PPP 是面向字符的，而 HDLC 是面向位的。

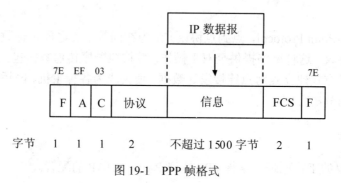

图 19-1  PPP 帧格式

可以看出，PPP 帧的前 3 个字段和最后 2 个字段与 HDLC 的格式是一样的。标志字段 F 为 0x7E（0x 表示 7E），但地址字段 A 和控制字段 C 都是固定不变的，分别为 0xFF 和 0x03。PPP 协议不是面向比特的，因而所有的 PPP 帧长度都是整数个字节。

PPP 帧与 HDLC 不同的是多了 2 个字节的协议字段。协议字段不同，后面的信息字段类型就不同。如：

0x0021——信息字段是 IP 数据报。

0xC021——信息字段是链路控制数据 LCP。

0x8021——信息字段是网络控制数据 NCP。

0xC023——信息字段是安全性认证 PAP。

0xC025——信息字段是 LQR。

0xC223——信息字段是安全性认证 CHAP。

当信息字段中出现和标志字段一样的比特 0x7E 时，就必须采取一些措施。因为 PPP 协议是面向字符型的，所以它不能采用 HDLC 所使用的零比特插入法，而是使用一种特殊的字符填充。具体的做法是将信息字段中出现的每一个 0x7E 字节转变成 2 字节序列（0x7D,0x5E）。若信息字段中出现一个 0x7D 的字节，则将其转变成 2 字节序列（0x7D,0x5D）。若信息字段中出现 ASCII 码的控制字符，则在该字符前面要加入一个 0x7D 字节。这样做的目的是防止这些表面上的 ASCII 码控制字符被错误地解释为控制字符。

(4) PPP 链路工作过程。

当用户拨号接入 ISP 时，路由器的调制解调器对拨号做出应答，并建立一条物理连接。这时 PC 机向路由器发送一系列的 LCP 分组（封装成多个 PPP 帧）。这些分组及其响应选择了将要使用的一些 PPP 参数，接着就进行网络层培植，NCP 给新接入的 PC 机分配一个临时的 IP 地址，这样 PC 机就成为 Internet 上的一个主机了。

当用户通信完毕时，NCP 释放网络层连接，收回原来分配出去的 IP 地址。接着 LCP 释放数据链路层连接，最后释放的是物理层的连接。

上述过程可用图 19-2 来描述。

当线路处于静止状态时，并不存在物理层的连接。当检测到调制解调器的载波信号，并建立物理层连接后，线路就进入建立状态，这时 LCP 开始协商一些选项。协商结束后就进入鉴别状态。若通信的双方鉴别身份成功，则进入网络状态。NCP 配置网络层，分配 IP 地址，然后就进入可进行数据通信的打开状态。数据传输结束后就转到终止状态。载波停止后则回到静止状态。

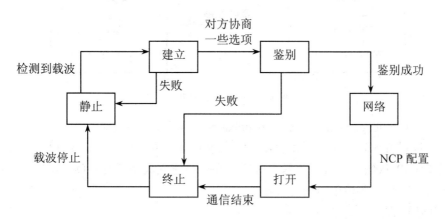

图 19-2　PPP 协议过程状态图

一个典型的链路建立过程分为三个阶段：创建阶段、认证阶段和网络协商阶段。

1）创建 PPP 链路。

LCP 负责创建链路。在这个阶段，将对基本的通信方式进行选择。链路两端设备通过 LCP 向对方发送配置信息报文（Configure Packets）。一旦一个配置成功信息包（Configure-Ack Packet）被发送且被接收，就完成了交换，进入了 LCP 开启状态。

2）用户验证。

在这个阶段，客户端会将自己的身份发送给远端的接入服务器。该阶段使用一种安全验证方式避免第三方窃取数据或冒充远程客户接管与客户端的连接。在认证完成之前，禁止从认证阶段前进到网络层协议阶段。如果认证失败，认证者应该跃迁到链路终止阶段。在这一阶段里，只有链路控制协议、认证协议和链路质量监视协议的 packets 是被允许的。在该阶段里接收到的其他 packets 必须被静静地丢弃。

3）调用网络层协议。

认证阶段完成之后，PPP 将调用在链路创建阶段选定的各种网络控制协议（NCP）。选定的 NCP 解决 PPP 链路之上的高层协议问题，例如，在该阶段，IP 控制协议（IPCP）可以向拨入用户分配动态地址。

这样，经过三个阶段以后，一条完整的 PPP 链路就建立起来了。

（5）PPP 协议的认证方式。

1）PAP（Password Authentication Protocol，口令认证协议）。

PAP 认证过程非常简单，采用二次握手机制；使用明文格式发送用户名和密码；认证发起方为被认证方可以做无限次的尝试（暴力破解）；只在链路建立的阶段进行 PAP 认证，一旦链路建立成功将不再进行认证检测。目前在 PPPOE 拨号环境中用得比较多。

PAP 认证过程：首先被认证方向主认证方发送认证请求（包含用户名和密码），主认证方接到认证请求，再根据被认证方发送来的用户名去到自己的数据库认证用户名密码是否正确，如果密码正确，PAP 认证通过；如果用户名密码错误，PAP 认证未通过。如图 19-3 所示。

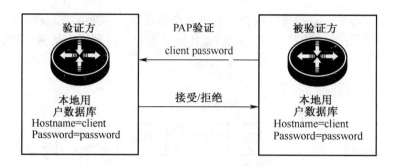

图 19-3  PAP 认证过程图

2）CHAP（Challenge Handshake Authentication Protocol，质询握手认证协议）。

CHAP 认证过程比较复杂，采用三次握手机制；使用密文格式发送 CHAP 认证信息；由认证方发起 CHAP 认证，有效避免暴力破解；在链路建立成功后具有再次认证检测机制。目前在企业网的远程接入环境中用得比较多。

CHAP 认证过程：

CHAP 认证第一步：主认证方发送挑战信息：01（此报文为认证请求）、ID（此认证的序列号）、随机数据、主认证方认证用户名；被认证方接收到挑战信息，根据接收到主认证方的认证用户名，到自己本地的数据库中查找对应的密码（如果没有设密码，就用默认的密码），查到密码再结合主认证方发来的 ID 和随机数据，根据 MD5 算法算出一个 Hash 值。如图 19-4 所示。

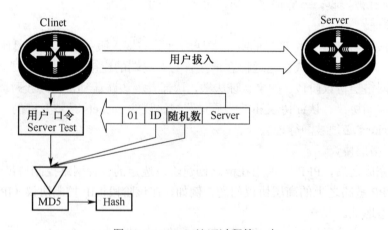

图 19-4  CHAP 认证过程第一步

CHAP 认证第二步：被认证方回复认证请求，认证请求里面包括：02（此报文为 CHAP 认证响应报文）、ID（与认证请求中的 ID 相同）、Hash 值、被认证方的认证用户名；主认证方处理挑战的响应信息，根据被认证方发来的认证用户名，主认证方在本地数据库中查找被认证方对应的密码（口令），结合 ID 找到先前保存的随机数据和 ID，根据 MD5 算法算出一个 Hash 值，与被认证方得到的 Hash 值做比较，如果一致，则认证通过，如果不一致，则认证不通过。如图 19-5 所示。

CHAP 认证第三步：认证方告知被认证方认证是否通过。如图 19-6 所示。

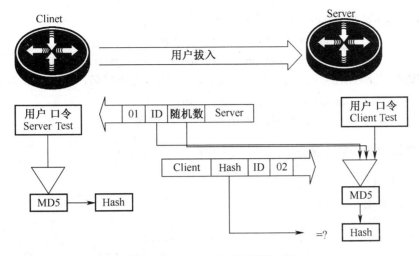

图 19-5　CHAP 认证过程第二步

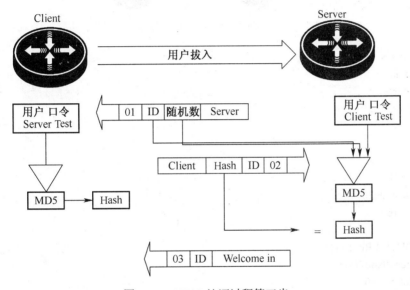

图 19-6　CHAP 认证过程第三步

## 四、实验需要掌握的命令

router(config-if)#encapsulation ppp
#设置端口封装的协议

router(config-if)#ppp authentication {chap | chap pap | pap chap | pap}
#设置认证方法

router(config-if)#clock rate {1200|2400|4800|9600|14400|19200|28800|38400|56000|64000|128000|2015232}
#设置 DCE 端线路速率

router(config-if)#ppp chap hostname *username*
#设置 CHAP 认证对端的认证用户名，*username* 为用户名

router(config-if)#ppp chap password *password*
#设置 CHAP 认证对端的认证密码，*password* 为密码

router(config-if)#ppp pap sent-username *username* password *password*
#设置 PAP 认证对端的认证用户名和密码，*username* 为用户名，*password* 为密码

router(config)#username *username* password *assword*
#创建一个用户，*username* 为用户名，*password* 为密码

### 五、实验拓扑图和拓扑文件

（1）网络拓扑图。

网络拓扑图如图 19-7 所示。

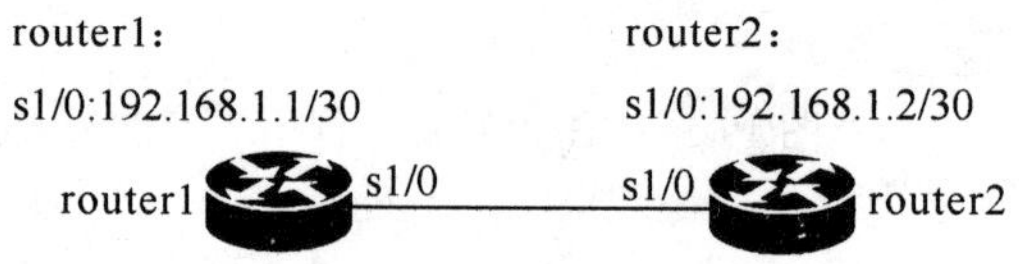

图 19-7　网络拓扑图

（2）网络文件。

```
autostart = false
ghostios = true
sparsemem = true

[localhost]

    [[3660]]
    image = ..\c3660-is.bin
    ram = 160

    [[ROUTER Router1]]
    idlepc = 0x60719b98
    model = 3660
    slot1 = NM-4T
    S1/0 = Router2 S1/0

    [[ROUTER Router2]]
    idlepc = 0x60719b98
    model = 3660
    slot1 = NM-4T
```

### 六、实验配置步骤

（1）没有认证的配置。

1）路由器 router1 的配置。

router>enable
router#configure terminal
router(config)#hostname R1

R1(config)#interface Serial1/0
R1(config-if)#ip address 192.168.1.1 255.255.255.252
R1(config-if)#encapsulation ppp
R1(config-if)#clock rate 64000
R1(config-if)#do write

2）路由器 router2 的配置。

router>enable
router#configure terminal
router(config)#hostname R2
R2(config)#interface Serial1/0
R2(config-if)#ip address 192.168.1.2 255.255.255.252
R2(config-if)#encapsulation ppp
R2(config-if)#clock rate 64000
R2(config-if)#do write

（2）PAP 认证的配置。

路由器 router1 的配置（在上面没有认证配置的基础上，增加下面的配置）
R1(config)#interface Serial1/0
R1(config-if)# ppp pap sent-username router2 password 0 a1b2c3
路由器 router2 的配置（在上面没有认证配置的基础上，增加下面的配置）
R2(config)#interface Serial1/0
R2(config)#username router2 password a1b2c3
R2(config-if)# ppp authentication pap

（3）CHAP 认证的配置。

路由器 router1 的配置（在上面没有认证配置的基础上，增加下面的配置）
R1(config)#interface Serial1/0
R1(config-if)#ppp chap hostname router2
R1(config-if)#ppp chap password 0 abc123
路由器 router2 的配置（在上面没有认证配置的基础上，增加下面的配置）
R2(config)#interface Serial1/0
R2(config)#username router2 password a1b2c3
R2(config-if)# ppp authentication chap

（4）CHAP 双向认证。

路由器 router1 的配置（在上面没有认证配置的基础上，增加下面的配置）
R1(config)#interface Serial1/0
R1(config)#username router1 password a1b2c3
R1(config-if)# ppp authentication chap
R1(config-if)#ppp chap hostname router2
R1(config-if)#ppp chap password 0 abc123
路由器 router2 的配置（在上面没有认证配置的基础上，增加下面的配置）
R2(config)#interface Serial1/0
R2(config)#username router2 password a1b2c3
R2(config-if)# ppp authentication chap
R2(config-if)#ppp chap hostname router2
R2(config-if)#ppp chap password 0 abc123

### 七、实验测试和查看相关配置

**（1）查看 router1 的端口状态。**

R1#show int s1/0
  Serial1/0 is up, line protocol is up　　#端口已经启用，端口线路协议已经开启
    Hardware is M4T
    Internet address is 192.168.1.1/30　　#端口的 IP 地址
    MTU 1500 bytes, BW 1544 Kbit, DLY 20000 usec,
      reliability 255/255, txload 1/255, rxload 1/255
    Encapsulation PPP, LCP Open　　#端口封装的协议为 PPP 协议，数据链路层控制协议（LCP）已经开启
    Open: IPCP, CDPCP, crc 16, loopback not set　　#网络层封装的协议是 IP 协议，应用 16 位的 CRC 进行检错，回环地址没有设置
    ......

**（2）查看 router2 的端口信息。**

R1#show int s1/0
  Serial1/0 is up, line protocol is up　　#端口已经启用，端口线路协议已经开启
    Hardware is M4T
    Internet address is 192.168.1.2/30　　#端口的 IP 地址
    MTU 1500 bytes, BW 1544 Kbit, DLY 20000 usec,
      reliability 255/255, txload 1/255, rxload 1/255
    Encapsulation PPP, LCP Open　　#端口封装的协议为 PPP 协议，数据链路层控制协议（LCP）已经开启
    Open: IPCP, CDPCP, crc 16, loopback not set　　#网络层封装的协议是 IP 协议，应用 16 位的 CRC 进行检错，回环地址没有设置
    ......

**（3）查看 router2 的 PAP 认证过程。**

R2#debug ppp authentication　　#开启 PPP 认证的调试开关
R2#sho debugging　　#显示调试信息
R2(config-if)#shutdown　　#关闭端口 s1/0
*Mar　1 01:34:27.555: %LINK-5-CHANGED: Interface Serial1/0, changed state to administratively down
*Mar　1 01:34:28.555: %LINEPROTO-5-UPDOWN: Line protocol on Interface Serial1/0, changed state to down
R2(config-if)#no shutdown　　#开启端口 s1/0
*Mar　1 02:11:47.471: %LINK-3-UPDOWN: Interface Serial1/0, changed state to up
*Mar　1 02:11:47.479: Se1/0 PPP: Using default call direction
*Mar　1 02:11:47.483: Se1/0 PPP: Treating connection as a dedicated line
*Mar　1 02:11:47.483: Se1/0 PPP: Session handle[6500000E] Session id[31]
*Mar　1 02:11:47.487: Se1/0 PPP: Authorization required
*Mar　1 02:11:47.679: Se1/0 PAP: I AUTH-REQ id 23 len 19 from "router2"
*Mar　1 02:11:47.683: Se1/0 PAP: Authenticating peer router2
*Mar　1 02:11:47.683: Se1/0 PPP: Sent PAP LOGIN Request
*Mar　1 02:11:47.691: Se1/0 PPP: Received LOGIN Response PASS
*Mar　1 02:11:47.699: Se1/0 PPP: Sent LCP AUTHOR Request
*Mar　1 02:11:47.699: Se1/0 PPP: Sent IPCP AUTHOR Request
*Mar　1 02:11:47.699: Se1/0 LCP: Received AAA AUTHOR Response PASS
*Mar　1 02:11:47.699: Se1/0 IPCP: Received AAA AUTHOR Response PASS
*Mar　1 02:11:47.703: Se1/0 PAP: O AUTH-ACK id 23 len 5
*Mar　1 02:11:47.711: Se1/0 PPP: Sent CDPCP AUTHOR Request

```
*Mar  1 02:11:47.723: Se1/0 PPP: Sent IPCP AUTHOR Request
*Mar  1 02:11:47.735: Se1/0 CDPCP: Received AAA AUTHOR Response PASS
*Mar  1 02:11:48.703: %LINEPROTO-5-UPDOWN: Line protocol on Interface Serial1/0, changed state to up
```

（4）查看 router2 的 CHAP 认证过程。

```
R2(config-if)#shutdown    #关闭端口 s1/0
*Mar  1 01:34:27.555: %LINK-5-CHANGED: Interface Serial1/0, changed state to administratively down
*Mar  1 01:34:28.555: %LINEPROTO-5-UPDOWN: Line protocol on Interface Serial1/0, changed state to down
R2(config-if)#no shutdown    #开启端口 s1/0
*Mar  1 02:13:53.447: %LINK-5-CHANGED: Interface Serial1/0, changed state to administratively down
*Mar  1 02:13:54.447: %LINEPROTO-5-UPDOWN: Line protocol on Interface Serial1/0, changed state to down
*Mar  1 02:13:55.583: %LINK-3-UPDOWN: Interface Serial1/0, changed state to up
*Mar  1 02:13:55.591: Se1/0 PPP: Using default call direction
*Mar  1 02:13:55.595: Se1/0 PPP: Treating connection as a dedicated line
*Mar  1 02:13:55.599: Se1/0 PPP: Session handle[D100000F] Session id[32]
*Mar  1 02:13:55.599: Se1/0 PPP: Authorization required
*Mar  1 02:13:55.699: Se1/0 CHAP: O CHALLENGE id 5 len 28 from "router2"
*Mar  1 02:13:55.751: Se1/0 CHAP: I RESPONSE id 5 len 28 from "router1"
*Mar  1 02:13:55.763: Se1/0 PPP: Sent CHAP LOGIN Request
*Mar  1 02:13:55.771: Se1/0 PPP: Received LOGIN Response PASS
*Mar  1 02:13:55.779: Se1/0 PPP: Sent LCP AUTHOR Request
*Mar  1 02:13:55.787: Se1/0 PPP: Sent IPCP AUTHOR Request
*Mar  1 02:13:55.795: Se1/0 LCP: Received AAA AUTHOR Response PASS
*Mar  1 02:13:55.803: Se1/0 IPCP: Received AAA AUTHOR Response PASS
*Mar  1 02:13:55.803: Se1/0 CHAP: O SUCCESS id 5 len 4
*Mar  1 02:13:55.811: Se1/0 PPP: Sent CDPCP AUTHOR Request
*Mar  1 02:13:55.823: Se1/0 CDPCP: Received AAA AUTHOR Response PASS
*Mar  1 02:13:55.883: Se1/0 PPP: Sent IPCP AUTHOR Request
*Mar  1 02:13:56.807: %LINEPROTO-5-UPDOWN: Line protocol on Interface Serial1/0, changed state to up
```

（5）查看 router2 的 CHAP 双向认证过程。

```
R2(config-if)#shutdown    #关闭端口 s1/0
*Mar  1 01:34:27.555: %LINK-5-CHANGED: Interface Serial1/0, changed state to administratively down
*Mar  1 01:34:28.555: %LINEPROTO-5-UPDOWN: Line protocol on Interface Serial1/0, changed state to down
R2(config-if)#no shutdown    #开启端口 s1/0
*Mar  1 01:34:34.195: %LINK-3-UPDOWN: Interface Serial1/0, changed state to up
*Mar  1 01:34:34.203: Se1/0 PPP: Using default call direction
*Mar  1 01:34:34.207: Se1/0 PPP: Treating connection as a dedicated line
*Mar  1 01:34:34.207: Se1/0 PPP: Session handle[49000008] Session id[24]
*Mar  1 01:34:34.211: Se1/0 PPP: Authorization required
*Mar  1 01:34:34.359: Se1/0 CHAP: O CHALLENGE id 4 len 28 from "router2"
*Mar  1 01:34:34.379: Se1/0 CHAP: I CHALLENGE id 5 len 28 from "router1"
*Mar  1 01:34:34.391: Se1/0 CHAP: Using hostname from interface CHAP
*Mar  1 01:34:34.395: Se1/0 CHAP: Using password from AAA
*Mar  1 01:34:34.395: Se1/0 CHAP: O RESPONSE id 5 len 28 from "router2"
*Mar  1 01:34:34.407: Se1/0 CHAP: I RESPONSE id 4 len 28 from "router1"
```

```
*Mar  1 01:34:34.419: Se1/0 PPP: Sent CHAP LOGIN Request
*Mar  1 01:34:34.427: Se1/0 PPP: Received LOGIN Response PASS
*Mar  1 01:34:34.435: Se1/0 PPP: Sent LCP AUTHOR Request
*Mar  1 01:34:34.435: Se1/0 PPP: Sent IPCP AUTHOR Request
*Mar  1 01:34:34.435: Se1/0 LCP: Received AAA AUTHOR Response PASS
*Mar  1 01:34:34.443: Se1/0 IPCP: Received AAA AUTHOR Response PASS
*Mar  1 01:34:34.443: Se1/0 CHAP: O SUCCESS id 4 len 4
*Mar  1 01:34:34.463: Se1/0 CHAP: I SUCCESS id 5 len 4
*Mar  1 01:34:34.471: Se1/0 PPP: Sent CDPCP AUTHOR Request
*Mar  1 01:34:34.487: Se1/0 PPP: Sent IPCP AUTHOR Request
*Mar  1 01:34:34.487: Se1/0 CDPCP: Received AAA AUTHOR Response PASS
*Mar  1 01:34:35.463: %LINEPROTO-5-UPDOWN: Line protocol on Interface Serial1/0, changed state to up
```

## 八、注意问题

区分认证方和被认证方，认证方设置认证的方式，被认证方设置认证用户名和密码。

# 实验二十　帧中继的配置

帧中继技术主要用于传递数据业务，它使用一组规程将数据信息以帧的形式（简称帧中继协议）有效地进行传送。它是广域网通信的一种方式；所使用的是逻辑连接，而不是物理连接，在一个物理连接上可复用多个逻辑连接，可实现带宽的复用和动态分配；帧中继协议是对 X.25 协议的简化，因此处理效率很高，网络吞吐量高，通信时延低，帧中继用户的接入速率在 64kbit/s 至 2Mbit/s，甚至可达到 34Mbit/s；帧中继的帧信息长度最大可达 1600 字节/帧，适合于封装局域网的数据单元及传送突发业务。

## 一、实验目的

（1）掌握帧中继的工作原理；
（2）熟练地配置帧中继。

## 二、实验内容

（1）帧中继交换机的配置；
（2）帧中继的 DLCI 的分配；
（3）接入帧中继网络的配置。

## 三、帧中继工作原理

帧中继（Frame Relay，FR）技术是在 OSI 第二层上，用简化的方法传送和交换数据单元的一种技术。帧中继技术是在分组技术充分发展、数字与光纤传输线路逐渐替代已有的模拟线路、用户终端日益智能化的条件下诞生并发展起来的。帧中继仅完成 OSI 物理层和链路层核心层的功能，将流量控制、纠错等留给智能终端去完成，大大简化了节点机之间的协议；同时，帧中继采用虚电路技术，能充分利用网络资源，因而帧中继具有吞吐量高、时延低、适合突发性业务等特点。作为一种新的承载业务，通过 RFC1490 协议，把网络层的 IP 数据包封装成数据链路层的帧中继帧，帧中继的用户接口速率最高为 34Mbit/s，目前在中、低速率网络互联的应用中被广泛使用。

帧中继技术适用于以下两种情况：

①用户需要数据通信，其带宽要求为 64kbit/s～34Mbit/s，而参与通信的各方多于两个的时候，使用帧中继是一种较好的解决方案。

②当数据业务量为突发性时，由于帧中继具有动态分配带宽的功能，选用帧中继可以有效地处理突发性数据。

（1）帧中继业务。

帧中继业务是在用户－网络接口（UNI）之间提供用户信息流的双向传送，并保持原顺序不变的一种承载业务。用户信息流以帧为单位在网络内传送，用户－网络接口之间以虚电路进行连接，对用户信息流进行统计复用。

帧中继网络提供的业务有两种：永久虚电路和交换虚电路。永久虚电路是指在帧中继终端用户之间建立固定的虚电路连接,并在其上提供数据传送业务。交换虚电路是指在数据传送前,两个帧中继终端用户之间通过呼叫建立虚电路连接,网络在建好的虚电路上提供数据信息的传送服务,终端用户通过呼叫清除操作终止虚电路。目前已建成的帧中继网络大多只提供永久虚电路业务。

帧中继永久虚电路业务模型如图 20-1 所示。

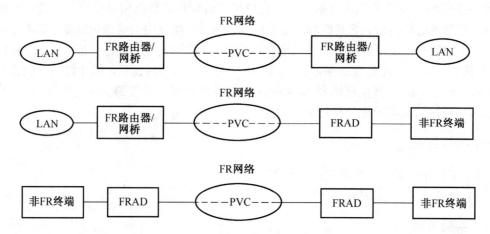

图 20-1 永久虚电路业务模型

(2) 帧中继的基本功能。

帧中继在 OSI 第二层以简化的方式传送数据,仅完成物理层和链路层核心层的功能,智能化的终端设备把数据发送到链路层,并封装在 LAPD 帧结构中,实施以帧为单位的信息传送。网络不进行纠错、重发、流量控制等。

帧不需要确认,就能够在每个交换机中直接通过,若网络检查出错误帧,直接将其丢弃;一些第二、三层的问题(如纠错、流量控制等)留给智能终端去处理,从而简化了节点机之间的处理过程。

帧中继承载业务有下列特点:

1) 全部控制平面的程序在逻辑上是分离的;

2) 物理层的用户平面程序使用 I.430/I.431 建议,链路层的用户平面程序使用 Q.922 建议的核心功能,能够对用户信息流量进行统计复用,并且可以保证在两个 S 或 T 参考点之间双向传送的业务数据单元的顺序。

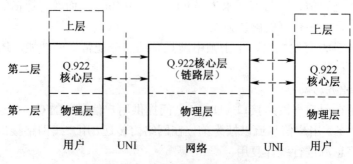

图 20-2 帧中继的协议结构

3）帧中继的帧结构。

帧中继的帧结构由 4 种字段组成，如图 20-3 所示。

| 1byte | 2～4byte | 1～4096byte | 2byte | 1byte |
|---|---|---|---|---|
| F | A（地址字段） | I（用户数据） | FCS | F |

图 20-3　帧中继的帧结构

F（标志字段）：由一个字节构成 01111110。它的作用是标志一个帧的开始和结束。为了防止在其他数据信息中随机出现的 01111110 序列影响同步，一般采用逢 5 插 1 的技术对数据进行处理，即对连续 5 个 1 位之后插入一个 0 位。在接收端再予以去除。

A（地址字段）：在帧中继中，地址字段的主要作用是寻址，同时还兼有拥塞管理功能。一般地址字段由 2 字节组成，如图 20-4 所示。如果 2 字节的地址字段不够用，因为 10 位的 DLCI 最多可支持 1023 个 PVC，如所需 PVC 数超过此限，则可扩展到 3 字节或 4 字节。目前我国未用。

| | 8　7　6　5 | 4 | 3 | 2 | 1 |
|---|---|---|---|---|---|
| byte1 | DLCI（高阶） | | | C/R | EA0 |
| byte2 | DLCI（低阶） | FECN | BECN | DE | EA1 |

图 20-4　2 字节 A 地址字段

DLCI：数据链路连接标识符，由 10 位构成，可提供 1023 个 PVC。

EA：扩充地址位，可将地址字段扩充到 3 或 4 字节，目前我国未用。最后一个字节的 EA 置 1，前面字节的 EA 置 0。

C/R：命令响应指示位，被透明地从一个终端传到另一个终端。它的用途是标识该帧是命令帧还是响应帧。命令帧的 C/R 位置 0，响应帧的 C/R 位置 1，目前我国未用。

FECN：前向显式拥塞通知，置 1 表示前向有可能发生拥塞。

BECN：后向显式拥塞通知，置 1 表示后向有可能发生拥塞。

DE：可丢弃位。DE 置 1，说明该帧在网络拥塞时可考虑丢弃。

I（用户数据）：信息字段，应由整数个字节组成。

FCS（帧校验序列）：一个 2 字节的序列，用于检验帧是否有差错。在帧中继网中，如传输产生差错，则该帧丢弃，由终端用户通知发端，重发此帧。

（4）DLCI 值的分配和帧中继的寻址机制。

对于 2 字节的 DLCI 可以有 1024 个取值，而对于 3 字节和 4 字节的 DLCI，则分别有 65536 个和 8388607 个取值。表 20-1 列出 2、3 和 4 字节的 DLCI 值的分配。

表 20-1　2、3 和 4 字节的 DLCI 值的分配

| 功能 | 2 字节 DLCI 值 | 3 字节 DLCI 值 | 4 字节 DLCI 值 |
|---|---|---|---|
| 通路内信令 | 0 | 0 | 0 |
| 保留 | 1～15 | 1～1023 | 1～1310711 |
| 用帧中继连接程序指定 | 16～991 | 1024～63487 | 131072～8126463 |
| 帧中继承载业务第二层管理 | 992～1007 | 63488～64511 | 8126464～8257535 |
| 保留 | 1008～1022 | 64512～65534 | 8257536～8388606 |
| 通路内层的管理 | 1023 | 65535 | 8388607 |

帧中继网中利用 DLCI 值完成寻址。而在帧中继的标准中有两种 DLCI 的定义方式。第一种方式：DLCI 值只有本地意义，即一个 DLCI 值只识别用户和接入交换节点，以及交换节点之间的逻辑连接。这样做的最大好处是相同 DLCI 值在整个帧中继网中可以重复使用。因此在确定的 DLCI 取值范围内能给出最大数目的虚电路连接。应注意的是，由于每段定义了不同的 DLCI 值，它只有本地的路由意义。第二种 DLCI 值的定义方式称为全网地址方式，此时 DLCI 值具有全网意义。它的优点是地址管理简单，为了增加可用的 DLCI 值，可将地址字段扩到 3 至 4 字节。

在帧中继网中，每一个交换节点内都有路由表存在。当用户需要建立端到端的永久虚电路连接（PVC）时，实际上是建立一条由多段 DLCI 连接而成的端到端的逻辑连接。当携带着用户数据信息的帧进入节点机后，根据该帧地址字段中的 DLCI 值，查找路由表以确定下一段逻辑连接的 DLCI 值，这样一站一站下去，直至目的地节点。

由于 DLCI 值是预先映射到目的节点上去的，这就简化了路由处理过程。节点机只需要关心路由表，在表中查看 DLCI 值，然后将业务内容路由到按照这个地址确定的输出端口上。在帧中继网中，一般不要求保持严格的 PVC 路由，可以是无连接的工作方式，允许在节点机间执行动态的路由，唯一的要求是确保帧到达由 DLCI 指定的目的端口。

图 20-5 是一个寻址的实例，阐明了在帧中继网中地址变换的过程。在这个例子中，终端用户在一个局域网上，路由器的 IP 地址为 128.1。终端用户设备发出的业务信息内容包含着目的网络地址 128.2 和目的主机地址 3.4。路由器 1 执行查表并决定将地址 128.2.3.4 映射到 DLCI 43。然后将用户业务内容放到帧中继的帧中，并将地址字段的 DLCI 值设为 43。交换节点 A 收到此帧，在路由表上将 DLCI 43 交换为 DLCI 76，并决定下一个接收此帧的节点为 B。节点 B 收到此帧后，将 DLCI 76 变换为 DLCI 84，并将此帧传递给路由器 2。路由器 2 查看到目的地址 128.2.3.4 后，将业务信息内容传递给终端用户 128.2.3.4。

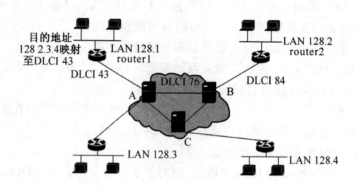

图 20-5　利用 DLCI 寻址的实例

## 四、实验需要掌握的命令

route(config)#frame-relay switching
#启用路由器模拟帧中继交换机

route(config-if)#encapsulation frame-relay
#在端口上封装帧中继协议

route(config-if)#frame-relay intf-type dce

#将端口设置为 DCE（数据传输设备）

route(config-if)#frame-relay intf-type dte

#将端口设置为 DTE（数据终端设备）（默认值）

route(config-if)#frame-relay route *DLCI-number* interface *type number DLCI-number*

#配置帧中继路由，第一个 *DLCI-number* 为输入端的帧中继号码，第二个帧中继号 *DLCI-number* 为输出端的帧中继号码；*type* 为接口的类型，*number* 为接口的编号

route(config-if)#frame-relay map ip *IP-address DLCI-number* broadcast

#将 IP 地址与帧中继号 *DLCI-number* 进行绑定

route(config-if)#frame-relay lmi-type cisco

#设置本地管理接口的类型

route(config)#show frame-relay pvc

#检查链路是否处于 active 状态

route(config)#show frame-relay map

#检查帧中继映射表

route(config)#show frame-relay lmi

#查看 lmi 信息

### 五、实验拓扑图和网络文件

（1）网络拓扑图。

网络拓扑图如图 20-6 所示。

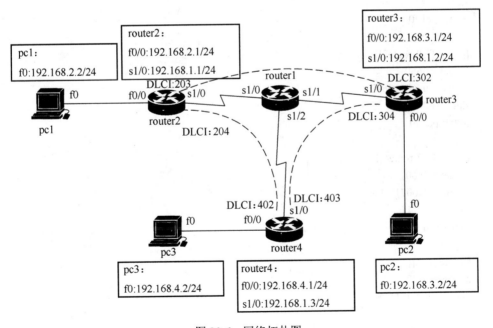

图 20-6 网络拓扑图

（2）网络文件。

#frame-relay.net 文件开始

autostart = false

ghostios = true
sparsemem = true

[localhost]

    [[3660]]
    image = ..\C3660-IS.BIN
    ram = 160

    [[1720]]
    image = ..\C1700-Y.BIN
    ram = 16

    [[ROUTER route1]]
    model = 3660
    idlepc = 0x60719b98
    slot1 = NM-4T
    s1/0 = route2 s1/0
    s1/1 = route3 s1/0
    s1/2 = route4 s1/0

    [[ROUTER route2]]
    model = 3660
    idlepc = 0x60719b98
    slot1 = NM-4T
    f0/0 = pc1 f0

    [[ROUTER route3]]
    model = 3660
    idlepc = 0x60719b98
    slot1 = NM-4T
    f0/0 = pc2 f0

    [[ROUTER route4]]
    model = 3660
    idlepc = 0x60719b98
    slot1 = NM-4T
    f0/0 = pc3 f0

    [[router pc1]]
    model = 1720
    idlepc = 0x8014e4ec

    [[router pc2]]
    model = 1720
    idlepc = 0x8014e4ec

    [[router pc3]]

        model = 1720
        idlepc = 0x8014e4ec
#frame-relay.net 文件结束

## 六、实验配置步骤

（1）模拟帧中继交换机的路由器 router1 的配置步骤。

router>enable
router#configure terminal
router(config)#hostname router1
router1(config)#frame-relay switching
router1(config)#interface Serial1/0
router1(config-if)#encapsulation frame-relay
router1(config-if)#clock rate 9600
router1(config-if)#frame-relay intf-type dce
router1(config-if)#frame-relay route 203 interface Serial1/1 302
router1(config-if)#frame-relay route 204 interface Serial1/2 402
router4(config-if)#no shutdown
router1(config-if)#interface Serial1/1
router1(config-if)#encapsulation frame-relay
router1(config-if)#clock rate 9600
router1(config-if)#frame-relay intf-type dce
router1(config-if)#frame-relay route 302 interface Serial1/0 203
router1(config-if)#frame-relay route 304 interface Serial1/2 403
router4(config-if)#no shutdown
router1(config-if)#interface Serial1/2
router1(config-if)#encapsulation frame-relay
router1(config-if)#clock rate 9600
router1(config-if)#frame-relay intf-type dce
router1(config-if)#frame-relay route 402 interface Serial1/0 204
router1(config-if)#frame-relay route 403 interface Serial1/1 304
router4(config-if)#no shutdown
router1(config-if)#do write

（2）路由器 router2 的配置步骤。

router>enable
router#configure terminal
router(config)#hostname router2
router2(config)#interface FastEthernet0/0
router2(config-if)#ip address 192.168.2.1 255.255.255.0
router2(config-if)#speed 100
router2(config-if)# duplex full
router4(config-if)#no shutdown
router2(config-if)#interface Serial1/0
router2(config-if)#ip address 192.168.1.1 255.255.255.0
router2(config-if)#encapsulation frame-relay
router2(config-if)#clock rate 9600
**router2(config-if)#frame-relay map ip 192.168.1.2 203 broadcast**
**router2(config-if)#frame-relay map ip 192.168.1.3 204 broadcast**

#上面加粗的两条命令也可以用下面加粗的命令代替
**router2(config-if)#frame-relay interface-dlci 203**
**router2(config-if)#frame-relay interface-dlci 204**
router4(config-if)#no shutdown
router4(config-if)#exit
router2(config)#ip route 192.168.3.0 255.255.255.0 192.168.1.2
router2(config)#ip route 192.168.4.0 255.255.255.0 192.168.1.3
#注意，此处的路由配置一定要用对端的 IP 地址，不可用本路由器的端口 s1/0 代替
router2(config)#do write

（3）路由器 router3 的配置步骤。

router>enable
router#configure terminal
router(config)#hostname router3
router3(config)#interface FastEthernet0/0
router3(config-if)#ip address 192.168.3.1 255.255.255.0
router3(config-if)#speed 100
router3(config-if)# duplex full
router4(config-if)#no shutdown
router3(config-if)#interface Serial1/0
router3(config-if)#ip address 192.168.1.2 255.255.255.0
router3(config-if)#encapsulation frame-relay
router3(config-if)#clock rate 9600
**router3(config-if)#frame-relay map ip 192.168.1.1 302 broadcast**
**router3(config-if)#frame-relay map ip 192.168.1.3 304 broadcast**
#上面加粗的两条命令也可以用下面加粗的命令代替
**router3(config-if)#frame-relay interface-dlci 302**
**router3(config-if)#frame-relay interface-dlci 304**
router4(config-if)#no shutdown
router3(config)#ip route 192.168.2.0 255.255.255.0 192.168.1.1
router3(config)#ip route 192.168.4.0 255.255.255.0 192.168.1.3
#注意，此处的路由配置一定要用对端的 IP 地址，不可用本路由器的端口 s1/0 代替
router3(config)#do write

（4）路由器 router4 的配置步骤。

router>enable
router#configure terminal
router(config)#hostname router4
router4(config)#interface FastEthernet0/0
router4(config-if)#ip address 192.168.4.1 255.255.255.0
router4(config-if)#speed 100
router4(config-if)#duplex full
router4(config-if)#no shutdown
router4(config-if)#interface Serial1/0
router4(config-if)#ip address 192.168.1.3 255.255.255.0
router4(config-if)#encapsulation frame-relay
router4(config-if)#clock rate 9600
**router4(config-if)#frame-relay map ip 192.168.1.1 402 broadcast**
**router4(config-if)#frame-relay map ip 192.168.1.2 403 broadcast**

#上面加粗的两条命令也可以用下面加粗的命令代替
**router4(config-if)#frame-relay interface-dlci 402**
**router4(config-if)#frame-relay interface-dlci 403**
router4(config-if)#no shutdown
router4(config-if)#exit
router4(config)#ip route 192.168.2.0 255.255.255.0 192.168.1.1
router4(config)#ip route 192.168.3.0 255.255.255.0 192.168.1.2
#注意，此处的路由配置一定要用对端的 IP 地址，不可用本路由器的端口 s1/0 代替
router4(config)#do write

(5) 主机 pc1 的配置步骤。

router>enable
router#configure terminal
router(config)#hostname pc1
pc1(config)#interface FastEthernet0
pc1(config-if)#ip address 192.168.2.2 255.255.255.0
pc1(config-if)#duplex full
pc1(config-if)#no shutdown
pc1(config-if)#exit
pc1(config)#no ip routing
pc1(config)#ip default-gateway 192.168.2.1
pc1(config)#do write

(6) 主机 pc2 的配置步骤。

router>enable
router#configure terminal
router(config)#hostname pc2
pc2(config)#interface FastEthernet0
pc2(config-if)#ip address 192.168.3.2 255.255.255.0
pc2(config-if)#duplex full
pc2(config-if)#no shutdown
pc2(config-if)#exit
pc2(config)#no ip routing
pc2(config)#ip default-gateway 192.168.3.1
pc2(config)#do write

(7) 主机 pc3 的配置步骤。

router>enable
router#configure terminal
router(config)#hostname pc3
pc3(config)#interface FastEthernet0
pc3(config-if)#ip address 192.168.4.2 255.255.255.0
pc3(config-if)#duplex full
pc3(config-if)#no shutdown
pc3(config-if)#exit
pc3(config)#no ip routing
pc3(config)#ip default-gateway 192.168.4.1
pc3(config)#do write

### 七、实验测试和查看相关配置

（1）查看帧中继交换机的帧中继路由。

router1#show frame-relay route

| Input Intf | Input Dlci | Output Intf | Output Dlci | Status |
|---|---|---|---|---|
| Serial1/0 | 203 | Serial1/1 | 302 | active |
| Serial1/0 | 204 | Serial1/2 | 402 | active |
| Serial1/1 | 302 | Serial1/0 | 203 | active |
| Serial1/1 | 304 | Serial1/2 | 403 | active |
| Serial1/2 | 402 | Serial1/0 | 204 | active |
| Serial1/2 | 403 | Serial1/1 | 304 | active |

（2）查看帧中继 PVC 详细信息。

router1#show frame-relay pvc 203
PVC Statistics for interface Serial1/0 (Frame Relay DCE)

DLCI = 203, DLCI USAGE = SWITCHED, PVC STATUS = ACTIVE, INTERFACE = Serial1/0

```
  input pkts 80          output pkts 70         in bytes 7816
  out bytes 7000         dropped pkts 0         in pkts dropped 0
  out pkts dropped 0            out bytes dropped 0
  in FECN pkts 0         in BECN pkts 0         out FECN pkts 0
  out BECN pkts 0        in DE pkts 0           out DE pkts 0
  out bcast pkts 0       out bcast bytes 0
  30 second input rate 0 bits/sec, 0 packets/sec
  30 second output rate 0 bits/sec, 0 packets/sec
  switched pkts 79
  Detailed packet drop counters:
  no out intf 0          out intf down 0        no out PVC 0
  in PVC down 0          out PVC down 0         pkt too big 0
  shaping Q full 0       pkt above DE 0         policing drop 0
  pvc create time 02:00:46, last time pvc status changed 01:45:55
```

（3）查看帧中继的 IP 以 DLCI 的映射。

router2#show frame-relay map
Serial1/0 (up): ip 192.168.1.2 dlci 203(0xCB,0x30B0), static,
                broadcast,
                CISCO, status defined, active
Serial1/0 (up): ip 192.168.1.3 dlci 204(0xCC,0x30C0), static,
                broadcast,
                CISCO, status defined, active

（4）测试网络的连通性。

pc1>ping 192.168.3.2
Type escape sequence to abort.
Sending 5, 100-byte ICMP Echos to 192.168.3.2, timeout is 2 seconds:
!!!!!
Success rate is 100 percent (5/5), round-trip min/avg/max = 84/208/328 ms
pc1>ping 192.168.4.2
Type escape sequence to abort.
Sending 5, 100-byte ICMP Echos to 192.168.4.2, timeout is 2 seconds:

!!!!!
Success rate is 100 percent (5/5), round-trip min/avg/max = 212/254/308 ms

## 八、注意问题

（1）两个直接连接的 Serial 端口的时钟速率一定要一致。

（2）两个直接连接的 Serial 端口封装的协议要一致，本实验都是 frame-relay。

（3）在设置模拟帧中继交换机的路由器的 Serial 端口时，端口的类型要设置为 DCE，默认的是 DTE。

（4）设置帧中继的路由时，要分清输入和输出的 DLCI 编号。

（5）在路由器上配置路由时，一定要用对端的 IP 地址，不可用本路由器的端口类型和端口号代替。

（6）在路由器上可以同时设置 frame-relay interface-dlci *DCLI-number* 和 frame-relay map ip *IP-address DCLI–number* broadcast。

（7）在用 frame-relay map ip *IP-address DCLI–number* broadcast 配置 IP 地址和 DLCI 号映射时要加上 broadcast 参数。

# 实验二十一  路由器配置综合实验

## 一、实验目的

通过综合实验的联系,深入了解路由交换知识体系,用所学知识点解决实际问题。

## 二、实验内容

依照拓扑图搭建环境,依照各部分需求达到全网互通。IP 地址自定义如下:
(1)交换部分:
①在四台交换机上创建 vlan1~vlan10;
②开启生成树协议;
③选定 sw5 为奇数 vlan 的根;sw6 为偶数 vlan 的根;
④在 sw5 与 sw6 之间做链路聚合(协议自定);
⑤sw8 的 f1/7 口做端口镜像,要求监控 f1/15 端口的入流量。
(2)路由部分:
①R2、R3、R4 属于 area 0;R3、R4、sw5 和 sw6 属于 area 1;
②其中 area 1 只能出现类型为 1、2、3 的 LSA,同时保证全网互通;
③R2 与 R3,R4 连接网段建立 OSPF 邻居关系,Hello 时间间隔为 10 秒;
④路由器 R3 和 R4 分别为 area 1 中相应网段的 DR;
⑤针对 sw7 和 sw8,在 sw5 和 sw6 上做 VRRP;
⑥R1 与 R2 建立 RIP 邻居关系,要求使用密文验证功能,密码为 a1b2c3。

## 三、实验拓扑图和拓扑文件

(1)网络拓扑图。
网络拓扑图如图 21-1 所示。

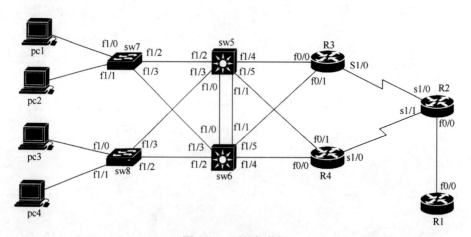

图 21-1  网络拓扑图

(2) 网络文件。
autostart = false
ghostios = true
sparsemem = true

[localhost]

　　[[3660]]
　　image = D:\Dynamips\images\c3660-is-mz.124-13.bin
　　ram = 128

　　[[1720]]
　　image = D:\Dynamips\images\C1700-Y.BIN
　　ram = 32

　　[[ROUTER R1]]
　　model = 3660
　　idlepc = 0x605d7170
　　f0/0 = R2 f0/0

　　[[ROUTER R2]]
　　model = 3660
　　idlepc = 0x605d7170
　　slot1 = NM-4T
　　s1/0 = R3 s1/0
　　s1/1 = R4 s1/0

　　[[ROUTER R3]]
　　model = 3660
　　slot1 = NM-4T
　　idlepc = 0x605d7170
　　f0/0 = sw5 f1/4
　　f0/1 = sw6 f1/5

　　[[ROUTER R4]]
　　model = 3660
　　slot1 = NM-4T
　　idlepc = 0x605d7170
　　f0/0 = sw6 f1/4
　　f0/1 = sw5 f1/5

　　[[ROUTER sw5]]
　　model = 3660
　　slot1 = NM-16ESW
　　idlepc = 0x605d7170
　　f1/0 = sw6 f1/0
　　f1/1 = sw6 f1/1
　　f1/2 = sw7 f1/2
　　f1/3 = sw8 f1/3

　　[[ROUTER sw6]]

model = 3660
slot1 = NM-16ESW
idlepc = 0x605d7170
f1/2 = sw8 f1/2
f1/3 = sw7 f1/3

[[ROUTER sw7]]
model = 3660
slot1 = NM-16ESW
idlepc = 0x605d7170
f1/0 = pc1 f0
f1/1 = pc2 f0

[[ROUTER sw8]]
model = 3660
slot1 = NM-16ESW
idlepc = 0x605d7170
f1/0 = pc3 f0
f1/1 = pc4 f0

[[ROUTER pc1]]
model = 1720
idlepc = 0x8014e4ec

[[ROUTER pc2]]
model = 1720
idlepc = 0x8014e4ec

[[ROUTER pc3]]
model = 1720
idlepc = 0x8014e4ec

[[ROUTER pc4]]
model = 1720
idlepc = 0x8014e4ec

## 四、IP 地址规划

设备端口地址规划如表 21-1 所示。

表 21-1　设备端口地址规划

| 设备名称 | 端口号 | IP 地址 | 备注 |
| --- | --- | --- | --- |
| R1 | f0/0 | 10.1.1.1/30 | |
| | loopback 0 | 192.168.1.1/32 | |
| R2 | f0/0 | 10.1.1.2/30 | |
| | s1/0 | 10.1.2.1/30 | |

续表

| 设备名称 | 端口号 | IP 地址 | 备注 |
| --- | --- | --- | --- |
| R2 | s1/1 | 10.1.3.1/30 | |
| | loopback 0 | 192.168.1.2/32 | |
| R3 | s1/0 | 10.1.2.2/30 | |
| | f0/0 | 10.1.4.1/30 | |
| | f0/1 | 10.1.4.9/30 | |
| | loopback 0 | 192.168.1.3/32 | |
| R4 | s1/0 | 10.1.3.2/30 | |
| | f0/0 | 10.1.4.13/30 | |
| | f0/1 | 10.1.4.5/30 | |
| | loopback 0 | 192.168.1.4/32 | |
| sw5 | f1/0 | | 链路聚合，配置 Trunk |
| | f1/1 | | |
| | f1/2 | | 配置 Trunk |
| | f1/3 | | 配置 Trunk |
| | f1/4 | 10.1.4.2/30 | |
| | f1/5 | 10.1.4.6/30 | |
| | vlan 1 | 10.2.1.2/24 | 配置 VRRP 组 1，虚拟 IP 地址为 10.2.1.1，模式为 Master |
| | vlan 2 | 10.2.2.2/24 | 配置 VRRP 组 2，虚拟 IP 地址为 10.2.2.1 |
| | vlan 3 | 10.2.3.2/24 | 配置 VRRP 组 3，虚拟 IP 地址为 10.2.3.1，模式为 Master |
| | vlan 4 | 10.2.4.2/24 | 配置 VRRP 组 4，虚拟 IP 地址为 10.2.4.1 |
| | vlan 5 | 10.2.5.2/24 | 配置 VRRP 组 5，虚拟 IP 地址为 10.2.5.1，模式为 Master |
| | vlan 6 | 10.2.6.2/24 | 配置 VRRP 组 6，虚拟 IP 地址为 10.2.6.1 |
| | vlan 7 | 10.2.7.2/24 | 配置 VRRP 组 7，虚拟 IP 地址为 10.2.7.1，模式为 Master |
| | vlan 8 | 10.2.8.2/24 | 配置 VRRP 组 8，虚拟 IP 地址为 10.2.8.1 |
| | vlan 9 | 10.2.9.2/24 | 配置 VRRP 组 9，虚拟 IP 地址为 10.2.9.1，模式为 Master |
| | vlan 10 | 10.2.10.2/24 | 配置 VRRP 组 10，虚拟 IP 地址为 10.2.10.1 |
| sw6 | f1/0 | | 链路聚合，配置 Trunk |
| | f1/1 | | |
| | f1/2 | | 配置 Trunk |
| | f1/3 | | 配置 Trunk |
| | f1/4 | 10.1.4.14/30 | |
| | f1/5 | 10.1.4.10/30 | |

续表

| 设备名称 | 端口号 | IP 地址 | 备注 |
|---|---|---|---|
| sw6 | vlan 1 | 10.2.1.3/24 | 配置 VRRP 组 1，虚拟 IP 地址为 10.2.1.1 |
| | vlan 2 | 10.2.2.3/24 | 配置 VRRP 组 2，虚拟 IP 地址为 10.2.2.1，模式为 Master |
| | vlan 3 | 10.2.3.3/24 | 配置 VRRP 组 3，虚拟 IP 地址为 10.2.3.1 |
| | vlan 4 | 10.2.4.3/24 | 配置 VRRP 组 4，虚拟 IP 地址为 10.2.4.1，模式为 Master |
| | vlan 5 | 10.2.5.3/24 | 配置 VRRP 组 5，虚拟 IP 地址为 10.2.5.1 |
| | vlan 6 | 10.2.6.3/24 | 配置 VRRP 组 6，虚拟 IP 地址为 10.2.6.1，模式为 Master |
| | vlan 7 | 10.2.7.3/24 | 配置 VRRP 组 7，虚拟 IP 地址为 10.2.7.1 |
| | vlan 8 | 10.2.8.3/24 | 配置 VRRP 组 8，虚拟 IP 地址为 10.2.8.1，模式为 Master |
| | vlan 9 | 10.2.9.3/24 | 配置 VRRP 组 9，虚拟 IP 地址为 10.2.9.1 |
| | vlan 10 | 10.2.10.3/24 | 配置 VRRP 组 10，虚拟 IP 地址为 10.2.10.1，模式为 Master |
| sw7 | f1/2 | | 配置 Trunk |
| | f1/3 | | 配置 Trunk |
| | f1/0 | | 属于 vlan2，连接到 pc1 |
| | f1/1 | | 属于 vlan3，连接到 pc2 |
| | f1/7 | | 配置端口镜像到端口 f1/15 |
| sw8 | f1/2 | | 配置 Trunk |
| | f1/3 | | 配置 Trunk |
| | f1/0 | | 属于 vlan2，连接到 pc3 |
| | f1/1 | | 属于 vlan3，连接到 pc4 |
| pc1 | f0 | 10.2.2.4/24 | 网关为 10.2.2.1 |
| pc2 | f0 | 10.2.3.4/24 | 网关为 10.2.3.1 |
| pc3 | f0 | 10.2.2.5/24 | 网关为 10.2.2.1 |
| pc4 | f0 | 10.2.3.5/24 | 网关为 10.2.3.1 |

### 五、实验配置步骤

（1）端口基本配置。

1）路由器 R1 的基本配置。

Router>enable
Router#config terminal
Router(config)#hostname R1
R1(config)#key chain router1
R1(config)#interface Loopback0
R1(config-if)#ip address 192.168.1.1 255.255.255.255

R1(config-if)#interface FastEthernet0/0
R1(config-if)#ip address 10.1.1.1 255.255.255.252
R1(config-if)#speed 100
R1(config-if)#duplex full
R1(config-if)#no shutdown
R1(config-if)#do write

2）路由器 R2 的基本配置。

Router>enable
Router#config terminal
Router(config)#hostname R2
R2(config)#interface Loopback0
R2(config-if)#ip address 192.168.1.2 255.255.255.255
R2(config-if)#interface FastEthernet0/0
R2(config-if)#ip address 10.1.1.2 255.255.255.252
R2(config-if)#speed 100
R2(config-if)#duplex full
R2(config-if)#no shutdown
R2(config-if)#interface Serial1/0
R2(config-if)#ip address 10.1.2.1 255.255.255.252
R2(config-if)#clock rate 64000
R2(config-if)#no shutdown
R2(config-if)#interface Serial1/1
R2(config-if)#ip address 10.1.3.1 255.255.255.252
R2(config-if)#clock rate 64000
R2(config-if)#no shutdown
R2(config-if)#do write

3）路由器 R3 的基本配置。

Router>enable
Router#config terminal
Router(config)#hostname R3
R3(config)#interface Loopback0
R3(config-if)#ip address 192.168.1.3 255.255.255.255
R3(config-if)#interface FastEthernet0/0
R3(config-if)#ip address 10.1.4.1 255.255.255.252
R3(config-if)#speed 100
R3(config-if)#duplex full
R3(config-if)#no shutdown
R3(config-if)#interface FastEthernet0/1
R3(config-if)#ip address 10.1.4.9 255.255.255.252
R3(config-if)#speed 100
R3(config-if)#duplex full
R3(config-if)#no shutdown
R3(config-if)#interface Serial1/0
R3(config-if)#ip address 10.1.2.2 255.255.255.252
R3(config-if)#clock rate 64000
R3(config-if)#no shutdown
R3 (config-if)#do write

4）路由器 R4 的基本配置。
Router>enable
Router#config terminal
Router(config)#hostname R4
R4(config)#interface Loopback0
R4(config-if)#ip address 192.168.1.4 255.255.255.255
R4(config-if)#interface FastEthernet0/0
R4(config-if)#ip address 10.1.4.13 255.255.255.252
R4(config-if)#speed 100
R4(config-if)#duplex full
R4(config-if)#no shutdown
R4(config-if)#interface FastEthernet0/1
R4(config-if)#ip address 10.1.4.5 255.255.255.252
R4(config-if)#speed 100
R4(config-if)#full-duplex
R4(config-if)#no shutdown
R4(config-if)#interface Serial1/0
R4(config-if)#ip address 10.1.3.2 255.255.255.252
R4(config-if)#clock rate 64000
R4(config-if)#no shutdown
R4(config-if)#do write

5）交换机 sw5 的基本配置。
Router>enable
Router#config terminal
Router(config)#hostname sw5
sw5 (config)#exit
sw5 #vlan database
sw5(vlan) #vlan 2
sw5(vlan) #vlan 3
sw5(vlan) #vlan 4
sw5(vlan) #vlan 5
sw5(vlan) #vlan 6
sw5(vlan) #vlan 7
sw5(vlan) #vlan 8
sw5(vlan) #vlan 9
sw5(vlan) #vlan 10
sw5(vlan) #exit
sw5#config terminal
sw5(config)#interface Loopback0
sw5(config-if)#ip address 192.168.1.5 255.255.255.255
sw5(config-if)#interface Vlan1
sw5(config-if)#ip address 10.2.1.2 255.255.255.0
sw5(config-if)#no shutdown
sw5(config-if)#interface Vlan2
sw5(config-if)#ip address 10.2.2.2 255.255.255.0
sw5(config-if)#no shutdown
sw5(config-if)#interface Vlan3

sw5config-if)#ip address 10.2.3.2 255.255.255.0
sw5(config-if)#no shutdown
sw5(config-if)#interface Vlan4
sw5(config-if)#ip address 10.2.4.2 255.255.255.0
sw5(config-if)#no shutdown
sw5(config-if)#interface Vlan5
sw5(config-if)#ip address 10.2.5.2 255.255.255.0
sw5(config-if)#no shutdown
sw5(config-if)#interface Vlan6
sw5(config-if)#ip address 10.2.6.2 255.255.255.0
sw5(config-if)#no shutdown
sw5(config-if)#interface Vlan7
sw5(config-if)#ip address 10.2.7.2 255.255.255.0
sw5(config-if)#no shutdown
sw5(config-if)#interface Vlan8
sw5(config-if)#ip address 10.2.8.2 255.255.255.0
sw5(config-if)#no shutdown
sw5(config-if)#interface Vlan9
sw5(config-if)#ip address 10.2.9.2 255.255.255.0
sw5(config-if)#no shutdown
sw5(config-if)#interface Vlan10
sw5(config-if)#ip address 10.2.10.2 255.255.255.0
sw5(config-if)#no shutdown
sw5(config-if)# interface FastEthernet1/4
sw5(config-if)#no switchport
sw5(config-if)#ip address 10.1.4.2 255.255.255.252
sw5(config-if)#no shutdown
sw5(config-if)# interface FastEthernet1/5
sw5(config-if)#no switchport
sw5(config-if)#ip address 10.1.4.6 255.255.255.252
sw5(config-if)#no shutdown
sw5(config-if)#do write

6）交换机 sw6 的基本配置。
Router>enable
Router#config terminal
Router(config)#hostname sw6
sw6(config)#exit
sw6#vlan database
sw6(vlan) #vlan 2
sw6(vlan) #vlan 3
sw6(vlan) #vlan 4
sw6(vlan) #vlan 5
sw6(vlan) #vlan 6
sw6(vlan) #vlan 7
sw6(vlan) #vlan 8
sw6(vlan) #vlan 9
sw6(vlan) #vlan 10

```
sw6(vlan) #exit
sw6#config terminal
sw6(config)#interface Loopback0
sw6(config-if)#ip address 192.168.1.6 255.255.255.255
sw6(config-if)#interface Vlan1
sw6(config-if)#ip address 10.2.1.3 255.255.255.0
sw6(config-if)#no shutdown
sw6(config-if)#interface Vlan2
sw6(config-if)#ip address 10.2.2.3 255.255.255.0
sw6(config-if)#no shutdown
sw6(config-if)#interface Vlan3
sw6(config-if)#ip address 10.2.3.3 255.255.255.0
sw6(config-if)#no shutdown
sw6(config-if)#interface Vlan4
sw6(config-if)#ip address 10.2.4.3 255.255.255.0
sw6(config-if)#no shutdown
sw6(config-if)#interface Vlan5
sw6(config-if)#ip address 10.2.5.3 255.255.255.0
sw6(config-if)#no shutdown
sw6(config-if)#interface Vlan6
sw6(config-if)#ip address 10.2.6.3 255.255.255.0
sw6(config-if)#no shutdown
sw6(config-if)#interface Vlan7
sw6(config-if)#ip address 10.2.7.3 255.255.255.0
sw6(config-if)#no shutdown
sw6(config-if)#interface Vlan8
sw6(config-if)#ip address 10.2.8.3 255.255.255.0
sw6(config-if)#no shutdown
sw6(config-if)#interface Vlan9
sw6(config-if)#ip address 10.2.9.3 255.255.255.0
sw6(config-if)#no shutdown
sw6(config-if)#interface Vlan10
sw6(config-if)#ip address 10.2.10.3 255.255.255.0
sw6(config-if)#no shutdown
sw6(config-if-range)#interface FastEthernet1/4
sw6(config-if)#no switchport
sw6(config-if)#ip address 10.1.4.14 255.255.255.252
sw6(config-if)#no shutdown
sw6(config-if)#interface FastEthernet1/5
sw6(config-if)#no switchport
sw6(config-if)#ip address 10.1.4.10 255.255.255.252
sw6(config-if)#no shutdown
sw6(config-if)#do write
```

7）交换机 sw7 的基本配置。

```
Router>enable
Router#config terminal
Router(config)#hostname sw7
```

sw7(config)#exit
sw7#vlan database
sw7(vlan) #vlan 2
sw7(vlan) #vlan 3
sw7(vlan) #vlan 4
sw7(vlan) #vlan 5
sw7(vlan) #vlan 6
sw7(vlan) #vlan 7
sw7(vlan) #vlan 8
sw7(vlan) #vlan 9
sw7(vlan) #vlan 10
sw7(vlan) #exit
sw7#write

8) 交换机 sw8 的基本配置。
Router>enable
Router#config terminal
Router(config)#hostname sw8
sw8(config)#exit
sw8#vlan database
sw8(vlan) #vlan 2
sw8(vlan) #vlan 3
sw8(vlan) #vlan 4
sw8(vlan) #vlan 5
sw8(vlan) #vlan 6
sw8(vlan) #vlan 7
sw8(vlan) #vlan 8
sw8(vlan) #vlan 9
sw8(vlan) #vlan 10
sw8(vlan) #exit
sw8#write

（2）根据实验要求对相关应用进行配置。
1) 路由器 R1 的应用配置。
R1(config)#key chain router1
R1(config-keychain)#key 1
R1(config-keychain-key)#key-string cisco
R1(config-keychain-key)#exit
R1(config-keychain)#exit
R1(config)#interface FastEthernet0/0
R1(config-if)#ip rip send version 2
R1(config-if)#ip rip receive version 2
R1(config-if)#ip rip authentication mode md5
R1(config-if)#ip rip authentication key-chain router1
#上面四条命令是配置 RIP 路由协议的认证
R1(config-if)#exit
R1(config)#router rip
R1(config-router)#version 2
R1(config-router)#network 10.0.0.0

R1(config-router)#default-information originate   #生成默认路由
R1(config-router)#no auto-summary   #关闭路由汇总
R1(config-router)#exit
R1(config)#ip route 0.0.0.0 0.0.0.0 FastEthernet0/1 permanent   #增加默认路由，使网络访问局域网以外的网络的路径
R1(config)#do write

2）路由器 R2 的应用配置。

R2(config)#key chain router2
R1(config-keychain)#key 1
R1(config-keychain-key)#key-string cisco
R1(config-keychain-key)#exit
R1(config-keychain)#exit
R2(config)#router rip
R2(config-router)#version 2
R2(config-router)#redistribute ospf 1 metric 4
R2(config-router)#network 10.0.0.0
R2(config-router)#no auto-summary
R2(config-router)#router ospf 1
R2(config-router)#redistribute rip subnets
R2(config-router)#network 10.1.2.0 0.0.0.3 area 0
R2(config-router)#network 10.1.3.0 0.0.0.3 area 0
R2(config-router)#network 192.168.1.2 0.0.0.0 area 0
R2(config-router)#default-information originate always
R2(config-router)#exit
R2(config)#interface FastEthernet0/0
R2(config-if)#ip rip send version 2
R2(config-if)#ip rip receive version 2
R2(config-if)#ip rip authentication mode md5
R2(config-if)#ip rip authentication key-chain router2
R2(config-if)#do write

3）路由器 R3 的应用配置。

R3(config)#router ospf 1
R3(config-router)#area 1 stub no-summary
R3(config-router)#network 10.1.2.0 0.0.0.3 area 0
R3(config-router)#network 10.1.4.0 0.0.0.3 area 1
R3(config-router)#network 10.1.4.8 0.0.0.3 area 1
R3(config-router)#network 192.168.1.3 0.0.0.0 area 0
R3(config)#interface FastEthernet0/0
R3(config-if)#ip ospf priority 100   #将此路由器设置 10.1.4.0/30 网段的 DR
R3(config-if)#interface FastEthernet0/1
R3(config-if)#ip ospf priority 100   #将此路由器设置 10.1.4.8/30 网段的 DR
R3(config-if)#do write

4）路由器 R4 的应用配置。

R4(config)#router ospf 1
R4(config-router)#area 1 stub no-summary
R4(config-router)#network 10.1.3.0 0.0.0.3 area 0
R4(config-router)#network 10.1.4.4 0.0.0.3 area 1

R4(config-router)#network 10.1.4.12 0.0.0.3 area 1
R4(config-router)#network 192.168.1.4 0.0.0.0 area 0
R4(config-router)#exit
R4(config)#interface FastEthernet0/0
R4(config-if)#ip ospf priority 100    #将此路由器设置 10.1.4.12/30 网段的 DR
R4(config-if)#interface FastEthernet0/1
R4(config-if)#ip ospf priority 100    #将此路由器设置 10.1.4.4/30 网段的 DR
R4(config-if)#do write

5）交换机 sw5 的应用配置。

sw5(config-if)#interface Port-channel1
sw5(config-if)#interface range FastEthernet1/0 - 1
sw5(config-if-range)#switchport mode trunk
sw5(config-if-range)#channel-group 1 mode on
sw5(config-if-range)#speed 100
sw5(config-if-range)#duplex full
sw5(config-if-range)#no shutdown
sw5(config-if-range)#interface Port-channel1
sw5(config-if)#switchport mode trunk
sw5(config-if)#interface range FastEthernet1/2 - 3
sw5(config-if-range)#switchport mode trunk
sw5(config-if-range)#speed 100
sw5(config-if-range)#duplex full
sw5(config-if-range)#no shutdown
sw5(config-if-range)#exit
sw5(config)#spanning-tree vlan 1 priority 8192
sw5(config)#spanning-tree vlan 3 priority 8192
sw5(config)#spanning-tree vlan 5 priority 8192
sw5(config)#spanning-tree vlan 7 priority 8192
sw5(config)#spanning-tree vlan 9 priority 8192
sw5(config)#track 1 interface FastEthernet1/4 line-protocol    #创建追踪列表 1，追踪端口 FastEthernet1/4 的线路协议是否为 up 状态
sw5(config-track)#exit
sw5(config)#track 2 interface FastEthernet1/5 line-protocol    #创建追踪列表 2，追踪端口 FastEthernet1/5 的线路协议是否为 up 状态
sw5(config-track)#exit
sw5(config)#track 3 list boolean or    #创建追踪列表 3，将追踪列表 1、2 中的端口状态进行或运算
sw5(config-track)#object 1    #定义目标，其中的 1 为 track 1 interface FastEthernet1/4 line-protocol 中的 1
sw5(config-track)#object 2    #定义目标，其中的 2 为 track 2 interface FastEthernet1/5 line-protocol 中的 2
sw5(config-track)#exit
sw5(config)#interface Vlan1
sw5(config-if)#vrrp 1 ip 10.2.1.1
sw5(config-if)#vrrp 1 priority 120
sw5(config-if)#vrrp 1 track 3 decrement 50
sw5(config-if)#interface Vlan2
sw5(config-if)#vrrp 2 ip 10.2.2.1
sw5(config-if)#interface Vlan3
sw5(config-if)#vrrp 3 ip 10.2.3.1

```
sw5(config-if)#vrrp 3 priority 120
sw5(config-if)#vrrp 3 track 3 decrement 50
sw5(config-if)#interface Vlan4
sw5(config-if)#vrrp 4 ip 10.2.4.1
sw5(config-if)#interface Vlan5
sw5(config-if)#vrrp 5 ip 10.2.5.1
sw5(config-if)#vrrp 5 priority 120
sw5(config-if)#vrrp 5 track 3 decrement 50
sw5(config-if)#interface Vlan6
sw5(config-if)#vrrp 6 ip 10.2.6.1
sw5(config-if)#interface Vlan7
sw5(config-if)#vrrp 7 ip 10.2.7.1
sw5(config-if)#vrrp 7 priority 120
sw5(config-if)#vrrp 7 track 3 decrement 50
sw5(config-if)#interface Vlan8
sw5(config-if)#vrrp 8 ip 10.2.8.1
sw5(config-if)#interface Vlan9
sw5(config-if)#vrrp 9 ip 10.2.9.1
sw5(config-if)#vrrp 9 priority 120
sw5(config-if)#vrrp 9 track 3 decrement 50
sw5(config-if)#interface Vlan10
sw5(config-if)#vrrp 10 ip 10.2.10.1
sw5(config-if)#exit
sw5(config)#router ospf 1
sw5(config-router)#area 1 stub no-summary
sw5(config-router)#network 10.1.4.0 0.0.0.7 area 1
sw5(config-router)#network 10.2.0.0 0.0.15.255 area 1
sw5(config-router)#network 192.168.1.5 0.0.0.0 area 1
sw5(config-router)#exit
sw5(config)#do write
```

6）交换机 sw6 的应用配置。

```
sw6(config-if)#interface Port-channel1
sw6(config-if)#interface range FastEthernet1/0 - 1
sw6(config-if-range)#switchport mode trunk
sw6(config-if-range)#channel-group 1 mode on
sw6(config-if-range)#speed 100
sw6(config-if-range)#duplex full
sw6(config-if-range)#no shutdown
sw6(config-if-range)#interface Port-channel1
sw6(config-if)#switchport mode trunk
sw6(config-if)#interface FastEthernet1/2 - 3
sw6(config-if-range)#switchport mode trunk
sw6(config-if-range)#speed 100
sw6(config-if-range)#duplex full
sw6(config-if-range)#no shutdown
sw6(config-if-range)#exit
sw6(config)#spanning-tree vlan 2 priority 8192
```

sw6(config)#spanning-tree vlan 4 priority 8192
sw6(config)#spanning-tree vlan 6 priority 8192
sw6(config)#spanning-tree vlan 8 priority 8192
sw6(config)#spanning-tree vlan 10 priority 8192
sw6(config)#track 1 interface FastEthernet1/4 line-protocol
sw6(config-track)#exit
sw6(config)#track 2 interface FastEthernet1/5 line-protocol
sw6(config-track)#exit
sw6(config)#track 3 list boolean or
sw6(config-track)#object 1
sw6(config-track)#object 2
#上面七条命令的解释见 sw5 的配置
sw6(config-track)#exit
sw6(config)#interface Vlan1
sw6(config-if)#vrrp 1 ip 10.2.1.1
sw6(config-if)#interface Vlan2
sw6(config-if)#vrrp 2 ip 10.2.2.1
sw6(config-if)#vrrp 2 priority 120
sw6(config-if)#vrrp 2 track 3 decrement 50
sw6(config-if)#interface Vlan3
sw6(config-if)#vrrp 3 ip 10.2.3.1
sw6(config-if)#interface Vlan4
sw6(config-if)#vrrp 4 ip 10.2.4.1
sw6(config-if)#vrrp 4 priority 120
sw6(config-if)#vrrp 4 track 3 decrement 50
sw6(config-if)#interface Vlan5
sw6(config-if)#vrrp 5 ip 10.2.5.1
sw6(config-if)#interface Vlan6
sw6(config-if)#vrrp 6 ip 10.2.6.1
sw6(config-if)#vrrp 6 priority 120
sw6(config-if)#vrrp 6 track 3 decrement 50
sw6(config-if)#interface Vlan7
sw6(config-if)#vrrp 7 ip 10.2.7.1
sw6(config-if)#interface Vlan8
sw6(config-if)# vrrp 8 ip 10.2.8.1
sw6(config-if)#vrrp 8 priority 120
sw6(config-if)#vrrp 8 track 3 decrement 50
sw6(config-if)#interface Vlan9
sw6(config-if)# vrrp 9 ip 10.2.9.1
sw6(config-if)#interface Vlan10
sw6(config-if)# vrrp 10 ip 10.2.10.1
sw6(config-if)#vrrp 10 priority 120
sw6(config-if)#vrrp 10 track 3 decrement 50
sw6(config-if)#exit
sw6(config)#router ospf 1
sw6(config-router)#area 1 stub no-summary
sw6(config-router)#network 10.1.4.8 0.0.0.7 area 1
sw6(config-router)#network 10.2.0.0 0.0.15.255 area 1

sw6(config-router)#network 192.168.1.6 0.0.0.0 area 1
sw6(config-router)#exit
sw6(config)#do write

7) 交换机 sw7 的应用配置。

R7(config)#interface range FastEthernet1/2 - 3
R7(config-if-range)#switchport mode trunk
R7(config-if-range)#interface FastEthernet1/0
R7(config-if)#switchport access vlan 2
R7(config-if)interface FastEthernet1/1
R7(config-if)#switchport access vlan 3
R7(config-if)#do write

8) 交换机 sw8 的应用配置。

R8(config)#monitor session 1 source interface Fa1/7 rx
R8(config)#monitor session 1 destination interface Fa1/15
R8(config)#interface range FastEthernet1/2 - 3
R8(config-if-range)#switchport mode trunk
R8(config-if-range)#interface FastEthernet1/0
R8(config-if)#switchport access vlan 2
R8(config-if)interface FastEthernet1/1
R8(config-if)#switchport access vlan 3
R8(config-if)#do write

此处省略了 pc1、pc2、pc3 和 pc4 的配置信息，参照前面的实验进行配置。

## 六、实验测试和查看相关配置

（1）基本测试。

测试每一个对接的端口是否能够正常通信，例如：R1#ping 10.1.1.2；如果通信不正常，首先要检查端口是否启动，例如：

R1#show interfaces f0/0
FastEthernet0/0 is up, line protocol is up
……

如果端口没有启动，进入端口配置模式，用 no shutdown 将端口启动，如果线路协议没有启起，则要查看对端的端口配置；其次看两端端口的速率是否一致（不要用 speed auto），如果不一致，就用命令 speed 100 将端口速率设为一致；第三查看两端端口通信方式是否一致，如果不一致，用命令 duplex 将两端端口设为一致，如果两端端口都支持全双工，最好都设置为全双工；第四要看端口 IP 地址是否在一个网段，同时要特别注意子网掩码是否一致，如果子网掩码不一致，通常会导致直接 ping 对端端口时是正常的，如果中间隔有其他网段或者通过其他路由器时，出现通信不正常。

检查二层端口的 Trunk 配置，例如：

Sw5#show interfaces f1/2 trunk

| Port | Mode | Encapsulation | Status | Native vlan |
|------|------|---------------|--------|-------------|
| Fa1/0 | on | 802.1q | trunking | 1 |

……

看 Trunk 协议是否正常启动了，如果没有启动，首先端口是否配置 Trunk 协议；其次查看端口是否启动；第三查看端口封装的协议是否一致。

（2）查看配置。

1）查看路由器 R1 的路由表信息。

R1# sho ip route

Codes: C - connected, S - static, R - RIP, M - mobile, B - BGP
　　　　D - EIGRP, EX - EIGRP external, O - OSPF, IA - OSPF inter area
　　　　N1 - OSPF NSSA external type 1, N2 - OSPF NSSA external type 2
　　　　E1 - OSPF external type 1, E2 - OSPF external type 2
　　　　i - IS-IS, su - IS-IS summary, L1 - IS-IS level-1, L2 - IS-IS level-2
　　　　ia - IS-IS inter area, * - candidate default, U - per-user static route
　　　　o - ODR, P - periodic downloaded static route

Gateway of last resort is 0.0.0.0 to network 0.0.0.0

　　　10.0.0.0/8 is variably subnetted, 18 subnets, 2 masks
R　　　10.2.8.0/24 [120/4] via 10.1.1.2, 00:00:14, FastEthernet0/0
R　　　10.2.9.0/24 [120/4] via 10.1.1.2, 00:00:14, FastEthernet0/0
R　　　10.2.10.0/24 [120/4] via 10.1.1.2, 00:00:14, FastEthernet0/0
R　　　10.1.4.12/30 [120/4] via 10.1.1.2, 00:00:14, FastEthernet0/0
R　　　10.1.4.8/30 [120/4] via 10.1.1.2, 00:00:14, FastEthernet0/0
R　　　10.1.3.0/30 [120/1] via 10.1.1.2, 00:00:14, FastEthernet0/0
R　　　10.2.1.0/24 [120/4] via 10.1.1.2, 00:00:14, FastEthernet0/0
R　　　10.1.2.0/30 [120/1] via 10.1.1.2, 00:00:14, FastEthernet0/0
R　　　10.2.2.0/24 [120/4] via 10.1.1.2, 00:00:14, FastEthernet0/0
C　　　10.1.1.0/30 is directly connected, FastEthernet0/0
R　　　10.2.3.0/24 [120/4] via 10.1.1.2, 00:00:14, FastEthernet0/0
R　　　10.1.4.4/30 [120/4] via 10.1.1.2, 00:00:14, FastEthernet0/0
R　　　10.2.4.0/24 [120/4] via 10.1.1.2, 00:00:15, FastEthernet0/0
R　　　10.2.5.0/24 [120/4] via 10.1.1.2, 00:00:15, FastEthernet0/0
R　　　10.2.6.0/24 [120/4] via 10.1.1.2, 00:00:15, FastEthernet0/0
C　　　10.1.1.4/30 is directly connected, FastEthernet0/1
R　　　10.2.7.0/24 [120/4] via 10.1.1.2, 00:00:15, FastEthernet0/0
R　　　10.1.4.0/30 [120/4] via 10.1.1.2, 00:00:15, FastEthernet0/0
　　　192.168.1.0/32 is subnetted, 6 subnets
C　　　192.168.1.1 is directly connected, Loopback0
R　　　192.168.1.3 [120/4] via 10.1.1.2, 00:00:15, FastEthernet0/0
R　　　192.168.1.2 [120/4] via 10.1.1.2, 00:00:15, FastEthernet0/0
R　　　192.168.1.5 [120/4] via 10.1.1.2, 00:00:15, FastEthernet0/0
R　　　192.168.1.4 [120/4] via 10.1.1.2, 00:00:15, FastEthernet0/0
R　　　192.168.1.6 [120/4] via 10.1.1.2, 00:00:15, FastEthernet0/0
S*　　 0.0.0.0/0 is directly connected, FastEthernet0/1

2）查看路由器 R2 的路由表信息。

R2#sho ip route

Codes: C - connected, S - static, R - RIP, M - mobile, B - BGP
　　　　D - EIGRP, EX - EIGRP external, O - OSPF, IA - OSPF inter area
　　　　N1 - OSPF NSSA external type 1, N2 - OSPF NSSA external type 2
　　　　E1 - OSPF external type 1, E2 - OSPF external type 2

```
       i - IS-IS, su - IS-IS summary, L1 - IS-IS level-1, L2 - IS-IS level-2
       ia - IS-IS inter area, * - candidate default, U - per-user static route
       o - ODR, P - periodic downloaded static route

Gateway of last resort is 10.1.1.1 to network 0.0.0.0

       10.0.0.0/8 is variably subnetted, 18 subnets, 2 masks
O IA    10.2.8.0/24 [110/66] via 10.1.3.2, 01:16:40, Serial1/1
                    [110/66] via 10.1.2.2, 01:16:40, Serial1/0
O IA    10.2.9.0/24 [110/66] via 10.1.3.2, 01:16:40, Serial1/1
                    [110/66] via 10.1.2.2, 01:16:40, Serial1/0
O IA    10.2.10.0/24 [110/66] via 10.1.3.2, 01:16:40, Serial1/1
                     [110/66] via 10.1.2.2, 01:16:40, Serial1/0
O IA    10.1.4.12/30 [110/65] via 10.1.3.2, 01:16:40, Serial1/1
O IA    10.1.4.8/30 [110/65] via 10.1.2.2, 01:16:37, Serial1/0
C       10.1.3.0/30 is directly connected, Serial1/1
O IA    10.2.1.0/24 [110/66] via 10.1.3.2, 01:16:40, Serial1/1
                    [110/66] via 10.1.2.2, 01:16:40, Serial1/0
C       10.1.2.0/30 is directly connected, Serial1/0
O IA    10.2.2.0/24 [110/66] via 10.1.3.2, 01:16:40, Serial1/1
                    [110/66] via 10.1.2.2, 01:16:40, Serial1/0
C       10.1.1.0/30 is directly connected, FastEthernet0/0
O IA    10.2.3.0/24 [110/66] via 10.1.3.2, 01:16:40, Serial1/1
                    [110/66] via 10.1.2.2, 01:16:40, Serial1/0
O IA    10.1.4.4/30 [110/65] via 10.1.3.2, 00:00:18, Serial1/1
O IA    10.2.4.0/24 [110/66] via 10.1.3.2, 01:16:40, Serial1/1
                    [110/66] via 10.1.2.2, 01:16:40, Serial1/0
O IA    10.2.5.0/24 [110/66] via 10.1.3.2, 01:16:40, Serial1/1
                    [110/66] via 10.1.2.2, 01:16:40, Serial1/0
O IA    10.2.6.0/24 [110/66] via 10.1.3.2, 01:16:40, Serial1/1
                    [110/66] via 10.1.2.2, 01:16:40, Serial1/0
R       10.1.1.4/30 [120/1] via 10.1.1.1, 00:00:01, FastEthernet0/0
O IA    10.2.7.0/24 [110/66] via 10.1.3.2, 01:16:40, Serial1/1
                    [110/66] via 10.1.2.2, 01:16:40, Serial1/0
O IA    10.1.4.0/30 [110/65] via 10.1.2.2, 00:00:28, Serial1/0
       192.168.1.0/32 is subnetted, 5 subnets
O       192.168.1.3 [110/65] via 10.1.2.2, 01:22:44, Serial1/0
C       192.168.1.2 is directly connected, Loopback0
O IA    192.168.1.5 [110/66] via 10.1.3.2, 00:00:18, Serial1/1
                    [110/66] via 10.1.2.2, 00:00:18, Serial1/0
O       192.168.1.4 [110/65] via 10.1.3.2, 01:22:44, Serial1/1
O IA    192.168.1.6 [110/66] via 10.1.3.2, 01:16:38, Serial1/1
                    [110/66] via 10.1.2.2, 01:16:38, Serial1/0
R*      0.0.0.0/0 [120/1] via 10.1.1.1, 00:00:01, FastEthernet0/0
```

3）查看路由器 R2 的 OSPF 的数据库。

R2#sho ip ospf database

OSPF Router with ID (192.168.1.2) (Process ID 1)

Router Link States (Area 0)

| Link ID | ADV Router | Age | Seq# | Checksum | Link count |
|---|---|---|---|---|---|
| 192.168.1.2 | 192.168.1.2 | 1012 | 0x80000007 | 0x003663 | 5 |
| 192.168.1.3 | 192.168.1.3 | 999 | 0x80000003 | 0x00D3EE | 3 |
| 192.168.1.4 | 192.168.1.4 | 1047 | 0x80000003 | 0x0008B5 | 3 |

Summary Net Link States (Area 0)

| Link ID | ADV Router | Age | Seq# | Checksum |
|---|---|---|---|---|
| 10.1.4.0 | 192.168.1.3 | 999 | 0x80000003 | 0x009D23 |
| 10.1.4.0 | 192.168.1.4 | 78 | 0x80000004 | 0x009F1E |
| 10.1.4.4 | 192.168.1.3 | 67 | 0x80000004 | 0x007D3D |
| 10.1.4.4 | 192.168.1.4 | 1047 | 0x80000003 | 0x006F4C |
| 10.1.4.8 | 192.168.1.3 | 999 | 0x80000003 | 0x004D6B |
| 10.1.4.8 | 192.168.1.4 | 788 | 0x80000003 | 0x005165 |
| 10.1.4.12 | 192.168.1.3 | 746 | 0x80000003 | 0x002F84 |
| 10.1.4.12 | 192.168.1.4 | 1047 | 0x80000003 | 0x001F94 |
| 10.2.1.0 | 192.168.1.3 | 746 | 0x80000003 | 0x00CEEF |
| 10.2.1.0 | 192.168.1.4 | 788 | 0x80000003 | 0x00C8F4 |
| 10.2.2.0 | 192.168.1.3 | 746 | 0x80000003 | 0x00C3F9 |
| 10.2.2.0 | 192.168.1.4 | 790 | 0x80000003 | 0x00BDFE |
| 10.2.3.0 | 192.168.1.3 | 748 | 0x80000003 | 0x00B804 |
| 10.2.3.0 | 192.168.1.4 | 790 | 0x80000003 | 0x00B209 |
| 10.2.4.0 | 192.168.1.3 | 748 | 0x80000003 | 0x00AD0E |
| 10.2.4.0 | 192.168.1.4 | 790 | 0x80000003 | 0x00A713 |
| 10.2.5.0 | 192.168.1.3 | 748 | 0x80000003 | 0x00A218 |
| 10.2.5.0 | 192.168.1.4 | 790 | 0x80000003 | 0x009C1D |
| 10.2.6.0 | 192.168.1.3 | 748 | 0x80000003 | 0x009722 |
| 10.2.6.0 | 192.168.1.4 | 790 | 0x80000003 | 0x009127 |
| 10.2.7.0 | 192.168.1.3 | 748 | 0x80000003 | 0x008C2C |
| 10.2.7.0 | 192.168.1.4 | 790 | 0x80000003 | 0x008631 |
| 10.2.8.0 | 192.168.1.3 | 748 | 0x80000003 | 0x008136 |
| 10.2.8.0 | 192.168.1.4 | 790 | 0x80000003 | 0x007B3B |
| 10.2.9.0 | 192.168.1.3 | 748 | 0x80000003 | 0x007640 |
| 10.2.9.0 | 192.168.1.4 | 790 | 0x80000003 | 0x007045 |
| 10.2.10.0 | 192.168.1.3 | 748 | 0x80000003 | 0x006B4A |
| 10.2.10.0 | 192.168.1.4 | 790 | 0x80000003 | 0x00654F |
| 192.168.1.5 | 192.168.1.3 | 68 | 0x80000005 | 0x0081D8 |
| 192.168.1.5 | 192.168.1.4 | 790 | 0x80000003 | 0x007FDB |
| 192.168.1.6 | 192.168.1.3 | 748 | 0x80000003 | 0x007BDF |
| 192.168.1.6 | 192.168.1.4 | 790 | 0x80000003 | 0x0075E4 |

Type-5 AS External Link States

| Link ID | ADV Router | Age | Seq# | Checksum | Tag |
|---|---|---|---|---|---|
| 0.0.0.0 | 192.168.1.2 | 1013 | 0x80000003 | 0x00182C | 1 |
| 10.1.1.0 | 192.168.1.2 | 1014 | 0x80000003 | 0x001910 | 0 |

10.1.1.4      192.168.1.2      1014      0x80000003      0x00F034 0

4）查看路由器 R3 的路由表信息。

R3#sho ip route
Codes: C - connected, S - static, R - RIP, M - mobile, B - BGP
       D - EIGRP, EX - EIGRP external, O - OSPF, IA - OSPF inter area
       N1 - OSPF NSSA external type 1, N2 - OSPF NSSA external type 2
       E1 - OSPF external type 1, E2 - OSPF external type 2
       i - IS-IS, su - IS-IS summary, L1 - IS-IS level-1, L2 - IS-IS level-2
       ia - IS-IS inter area, * - candidate default, U - per-user static route
       o - ODR, P - periodic downloaded static route

Gateway of last resort is 10.1.2.1 to network 0.0.0.0

       10.0.0.0/8 is variably subnetted, 18 subnets, 2 masks
O         10.2.8.0/24 [110/2] via 10.1.4.10, 00:01:40, FastEthernet0/1
                     [110/2] via 10.1.4.2, 00:01:40, FastEthernet0/0
O         10.2.9.0/24 [110/2] via 10.1.4.10, 00:01:40, FastEthernet0/1
                     [110/2] via 10.1.4.2, 00:01:40, FastEthernet0/0
O         10.2.10.0/24 [110/2] via 10.1.4.10, 00:01:40, FastEthernet0/1
                      [110/2] via 10.1.4.2, 00:01:40, FastEthernet0/0
O         10.1.4.12/30 [110/2] via 10.1.4.10, 00:01:40, FastEthernet0/1
C         10.1.4.8/30 is directly connected, FastEthernet0/1
O         10.1.3.0/30 [110/128] via 10.1.2.1, 01:24:03, Serial1/0
O         10.2.1.0/24 [110/2] via 10.1.4.10, 00:01:40, FastEthernet0/1
                     [110/2] via 10.1.4.2, 00:01:40, FastEthernet0/0
C         10.1.2.0/30 is directly connected, Serial1/0
O         10.2.2.0/24 [110/2] via 10.1.4.10, 00:01:43, FastEthernet0/1
                     [110/2] via 10.1.4.2, 00:01:43, FastEthernet0/0
O E2      10.1.1.0/30 [110/20] via 10.1.2.1, 00:01:43, Serial1/0
O         10.2.3.0/24 [110/2] via 10.1.4.10, 00:01:43, FastEthernet0/1
                     [110/2] via 10.1.4.2, 00:01:43, FastEthernet0/0
O         10.1.4.4/30 [110/2] via 10.1.4.2, 00:01:43, FastEthernet0/0
O         10.2.4.0/24 [110/2] via 10.1.4.10, 00:01:43, FastEthernet0/1
                     [110/2] via 10.1.4.2, 00:01:43, FastEthernet0/0
O         10.2.5.0/24 [110/2] via 10.1.4.10, 00:01:43, FastEthernet0/1
                     [110/2] via 10.1.4.2, 00:01:43, FastEthernet0/0
O         10.2.6.0/24 [110/2] via 10.1.4.10, 00:01:43, FastEthernet0/1
                     [110/2] via 10.1.4.2, 00:01:43, FastEthernet0/0
O E2      10.1.1.4/30 [110/20] via 10.1.2.1, 00:01:43, Serial1/0
O         10.2.7.0/24 [110/2] via 10.1.4.10, 00:01:43, FastEthernet0/1
                     [110/2] via 10.1.4.2, 00:01:43, FastEthernet0/0
C         10.1.4.0/30 is directly connected, FastEthernet0/0
       192.168.1.0/32 is subnetted, 5 subnets
C         192.168.1.3 is directly connected, Loopback0
O         192.168.1.2 [110/65] via 10.1.2.1, 01:24:06, Serial1/0
O         192.168.1.5 [110/2] via 10.1.4.2, 00:01:43, FastEthernet0/0
O         192.168.1.4 [110/129] via 10.1.2.1, 01:24:06, Serial1/0

O        192.168.1.6 [110/2] via 10.1.4.10, 00:01:43, FastEthernet0/1
O*E2 0.0.0.0/0 [110/1] via 10.1.2.1, 00:01:43, Serial1/0

5）查看路由器 R3 的 OSPF 的邻居信息。

R3#show ip ospf neighbor

| Neighbor ID | Pri | State | Dead Time | Address | Interface |
|---|---|---|---|---|---|
| 192.168.1.2 | 0 | FULL/ - | 00:00:36 | 10.1.2.1 | Serial1/0 |
| 192.168.1.6 | 1 | FULL/BDR | 00:00:37 | 10.1.4.10 | FastEthernet0/1 |
| 192.168.1.5 | 1 | FULL/BDR | 00:00:31 | 10.1.4.2 | FastEthernet0/0 |

6）查看路由器 R3 的 OSPF 数据库。

R3#show ip ospf database

OSPF Router with ID (192.168.1.3) (Process ID 1)

Router Link States (Area 0)

| Link ID | ADV Router | Age | Seq# | Checksum | Link count |
|---|---|---|---|---|---|
| 192.168.1.2 | 192.168.1.2 | 1129 | 0x80000007 | 0x003663 | 5 |
| 192.168.1.3 | 192.168.1.3 | 1114 | 0x80000003 | 0x00D3EE | 3 |
| 192.168.1.4 | 192.168.1.4 | 1164 | 0x80000003 | 0x0008B5 | 3 |

Summary Net Link States (Area 0)

| Link ID | ADV Router | Age | Seq# | Checksum |
|---|---|---|---|---|
| 10.1.4.0 | 192.168.1.3 | 1114 | 0x80000003 | 0x009D23 |
| 10.1.4.0 | 192.168.1.4 | 194 | 0x80000004 | 0x009F1E |
| 10.1.4.4 | 192.168.1.3 | 182 | 0x80000004 | 0x007D3D |
| 10.1.4.4 | 192.168.1.4 | 1164 | 0x80000003 | 0x006F4C |
| 10.1.4.8 | 192.168.1.3 | 1114 | 0x80000003 | 0x004D6B |
| 10.1.4.8 | 192.168.1.4 | 905 | 0x80000003 | 0x005165 |
| 10.1.4.12 | 192.168.1.3 | 861 | 0x80000003 | 0x002F84 |
| 10.1.4.12 | 192.168.1.4 | 1164 | 0x80000003 | 0x001F94 |
| 10.2.1.0 | 192.168.1.3 | 861 | 0x80000003 | 0x00CEEF |
| 10.2.1.0 | 192.168.1.4 | 905 | 0x80000003 | 0x00C8F4 |
| 10.2.2.0 | 192.168.1.3 | 861 | 0x80000003 | 0x00C3F9 |
| 10.2.2.0 | 192.168.1.4 | 905 | 0x80000003 | 0x00BDFE |
| 10.2.3.0 | 192.168.1.3 | 861 | 0x80000003 | 0x00B804 |
| 10.2.3.0 | 192.168.1.4 | 905 | 0x80000003 | 0x00B209 |
| 10.2.4.0 | 192.168.1.3 | 861 | 0x80000003 | 0x00AD0E |
| 10.2.4.0 | 192.168.1.4 | 905 | 0x80000003 | 0x00A713 |
| 10.2.5.0 | 192.168.1.3 | 861 | 0x80000003 | 0x00A218 |
| 10.2.5.0 | 192.168.1.4 | 905 | 0x80000003 | 0x009C1D |
| 10.2.6.0 | 192.168.1.3 | 861 | 0x80000003 | 0x009722 |
| 10.2.6.0 | 192.168.1.4 | 905 | 0x80000003 | 0x009127 |
| 10.2.7.0 | 192.168.1.3 | 861 | 0x80000003 | 0x008C2C |
| 10.2.7.0 | 192.168.1.4 | 905 | 0x80000003 | 0x008631 |
| 10.2.8.0 | 192.168.1.3 | 861 | 0x80000003 | 0x008136 |
| 10.2.8.0 | 192.168.1.4 | 905 | 0x80000003 | 0x007B3B |

| | | | | | |
|---|---|---|---|---|---|
| 10.2.9.0 | 192.168.1.3 | 861 | 0x80000003 | 0x007640 | |
| 10.2.9.0 | 192.168.1.4 | 905 | 0x80000003 | 0x007045 | |
| 10.2.10.0 | 192.168.1.3 | 861 | 0x80000003 | 0x006B4A | |
| 10.2.10.0 | 192.168.1.4 | 905 | 0x80000003 | 0x00654F | |
| 192.168.1.5 | 192.168.1.3 | 182 | 0x80000005 | 0x0081D8 | |
| 192.168.1.5 | 192.168.1.4 | 905 | 0x80000003 | 0x007FDB | |
| 192.168.1.6 | 192.168.1.3 | 862 | 0x80000003 | 0x007BDF | |
| 192.168.1.6 | 192.168.1.4 | 905 | 0x80000003 | 0x0075E4 | |

Router Link States (Area 1)

| Link ID | ADV Router | Age | Seq# | Checksum | Link count |
|---|---|---|---|---|---|
| 192.168.1.3 | 192.168.1.3 | 189 | 0x80000005 | 0x002CC8 | 2 |
| 192.168.1.4 | 192.168.1.4 | 906 | 0x80000005 | 0x003DA5 | 2 |
| 192.168.1.5 | 192.168.1.5 | 191 | 0x80000007 | 0x009AB2 | 13 |
| 192.168.1.6 | 192.168.1.6 | 783 | 0x80000005 | 0x00EE33 | 13 |

Net Link States (Area 1)

| Link ID | ADV Router | Age | Seq# | Checksum |
|---|---|---|---|---|
| 10.1.4.1 | 192.168.1.3 | 189 | 0x80000001 | 0x00F6ED |
| 10.1.4.5 | 192.168.1.4 | 905 | 0x80000003 | 0x00CE0E |
| 10.1.4.9 | 192.168.1.3 | 862 | 0x80000003 | 0x00B029 |
| 10.1.4.13 | 192.168.1.4 | 905 | 0x80000003 | 0x008C47 |
| 10.2.1.2 | 192.168.1.5 | 732 | 0x80000003 | 0x0026B5 |
| 10.2.2.2 | 192.168.1.5 | 732 | 0x80000003 | 0x001BBF |
| 10.2.3.2 | 192.168.1.5 | 732 | 0x80000003 | 0x0010C9 |
| 10.2.4.2 | 192.168.1.5 | 732 | 0x80000003 | 0x0005D3 |
| 10.2.5.2 | 192.168.1.5 | 732 | 0x80000003 | 0x00F9DD |
| 10.2.6.2 | 192.168.1.5 | 732 | 0x80000003 | 0x00EEE7 |
| 10.2.7.2 | 192.168.1.5 | 732 | 0x80000003 | 0x00E3F1 |
| 10.2.8.2 | 192.168.1.5 | 732 | 0x80000003 | 0x00D8FB |
| 10.2.9.2 | 192.168.1.5 | 732 | 0x80000003 | 0x00CD06 |
| 10.2.10.2 | 192.168.1.5 | 732 | 0x80000003 | 0x00C210 |

Summary Net Link States (Area 1)

| Link ID | ADV Router | Age | Seq# | Checksum |
|---|---|---|---|---|
| 0.0.0.0 | 192.168.1.3 | 1115 | 0x80000003 | 0x008846 |
| 0.0.0.0 | 192.168.1.4 | 1165 | 0x80000003 | 0x00824B |

Type-5 AS External Link States

| Link ID | ADV Router | Age | Seq# | Checksum | Tag |
|---|---|---|---|---|---|
| 0.0.0.0 | 192.168.1.2 | 1130 | 0x80000003 | 0x00182C | 1 |
| 10.1.1.0 | 192.168.1.2 | 1130 | 0x80000003 | 0x001910 | 0 |
| 10.1.1.4 | 192.168.1.2 | 1130 | 0x80000003 | 0x00F034 | 0 |

7）查看路由器 R4 的路由表信息。

R4#sho ip route
Codes: C - connected, S - static, R - RIP, M - mobile, B - BGP
   D - EIGRP, EX - EIGRP external, O - OSPF, IA - OSPF inter area
   N1 - OSPF NSSA external type 1, N2 - OSPF NSSA external type 2
   E1 - OSPF external type 1, E2 - OSPF external type 2
   i - IS-IS, su - IS-IS summary, L1 - IS-IS level-1, L2 - IS-IS level-2
   ia - IS-IS inter area, * - candidate default, U - per-user static route
   o - ODR, P - periodic downloaded static route

Gateway of last resort is not set

```
     10.0.0.0/8 is variably subnetted, 18 subnets, 2 masks
O       10.2.8.0/24 [110/2] via 10.1.4.14, 00:03:34, FastEthernet0/0
                    [110/2] via 10.1.4.6, 00:03:34, FastEthernet0/1
O       10.2.9.0/24 [110/2] via 10.1.4.14, 00:03:34, FastEthernet0/0
                    [110/2] via 10.1.4.6, 00:03:34, FastEthernet0/1
O       10.2.10.0/24 [110/2] via 10.1.4.14, 00:03:34, FastEthernet0/0
                     [110/2] via 10.1.4.6, 00:03:34, FastEthernet0/1
C       10.1.4.12/30 is directly connected, FastEthernet0/0
O       10.1.4.8/30 [110/2] via 10.1.4.14, 00:03:34, FastEthernet0/0
C       10.1.3.0/30 is directly connected, Serial1/0
O       10.2.1.0/24 [110/2] via 10.1.4.14, 00:03:34, FastEthernet0/0
                    [110/2] via 10.1.4.6, 00:03:34, FastEthernet0/1
O       10.1.2.0/30 [110/128] via 10.1.3.1, 01:25:59, Serial1/0
O       10.2.2.0/24 [110/2] via 10.1.4.14, 00:03:34, FastEthernet0/0
                    [110/2] via 10.1.4.6, 00:03:34, FastEthernet0/1
O E2    10.1.1.0/30 [110/20] via 10.1.3.1, 00:03:34, Serial1/0
O       10.2.3.0/24 [110/2] via 10.1.4.14, 00:03:34, FastEthernet0/0
                    [110/2] via 10.1.4.6, 00:03:34, FastEthernet0/1
C       10.1.4.4/30 is directly connected, FastEthernet0/1
O       10.2.4.0/24 [110/2] via 10.1.4.14, 00:03:34, FastEthernet0/0
                    [110/2] via 10.1.4.6, 00:03:34, FastEthernet0/1
O       10.2.5.0/24 [110/2] via 10.1.4.14, 00:03:34, FastEthernet0/0
                    [110/2] via 10.1.4.6, 00:03:34, FastEthernet0/1
O       10.2.6.0/24 [110/2] via 10.1.4.14, 00:03:34, FastEthernet0/0
                    [110/2] via 10.1.4.6, 00:03:34, FastEthernet0/1
O E2    10.1.1.4/30 [110/20] via 10.1.3.1, 00:03:34, Serial1/0
O       10.2.7.0/24 [110/2] via 10.1.4.14, 00:03:34, FastEthernet0/0
                    [110/2] via 10.1.4.6, 00:03:34, FastEthernet0/1
O       10.1.4.0/30 [110/2] via 10.1.4.6, 00:03:34, FastEthernet0/1
     192.168.1.0/32 is subnetted, 5 subnets
O       192.168.1.3 [110/129] via 10.1.3.1, 01:25:59, Serial1/0
O       192.168.1.2 [110/65] via 10.1.3.1, 01:25:59, Serial1/0
O       192.168.1.5 [110/2] via 10.1.4.6, 00:03:34, FastEthernet0/1
C       192.168.1.4 is directly connected, Loopback0
O       192.168.1.6 [110/2] via 10.1.4.14, 00:03:34, FastEthernet0/0
O*E2 0.0.0.0/0 [110/1] via 10.1.3.1, 00:03:34, Serial1/0
```

8）查看路由器 R4 的 OSPF 的邻居信息。

R4#sho ip ospf neighbor

| | | | | | | |
|---|---|---|---|---|---|---|
| 192.168.1.2 | 0 | FULL/ - | 00:00:38 | 10.1.3.1 | | Serial1/0 |
| 192.168.1.6 | 1 | FULL/BDR | 00:00:38 | 10.1.4.14 | | FastEthernet0/0 |
| 192.168.1.5 | 1 | FULL/BDR | 00:00:39 | 10.1.4.6 | | FastEthernet0/1 |

9）查看路由器 R4 的 OSPF 的数据库。

R4#sho ip ospf database

OSPF Router with ID (192.168.1.4) (Process ID 1)

Router Link States (Area 0)

| Link ID | ADV Router | Age | Seq# | Checksum | Link count |
|---|---|---|---|---|---|
| 192.168.1.2 | 192.168.1.2 | 1239 | 0x80000007 | 0x003663 | 5 |
| 192.168.1.3 | 192.168.1.3 | 1226 | 0x80000003 | 0x00D3EE | 3 |
| 192.168.1.4 | 192.168.1.4 | 1272 | 0x80000003 | 0x0008B5 | 3 |

Summary Net Link States (Area 0)

| Link ID | ADV Router | Age | Seq# | Checksum |
|---|---|---|---|---|
| 10.1.4.0 | 192.168.1.3 | 1226 | 0x80000003 | 0x009D23 |
| 10.1.4.0 | 192.168.1.4 | 302 | 0x80000004 | 0x009F1E |
| 10.1.4.4 | 192.168.1.3 | 294 | 0x80000004 | 0x007D3D |
| 10.1.4.4 | 192.168.1.4 | 1272 | 0x80000003 | 0x006F4C |
| 10.1.4.8 | 192.168.1.3 | 1226 | 0x80000003 | 0x004D6B |
| 10.1.4.8 | 192.168.1.4 | 1013 | 0x80000003 | 0x005165 |
| 10.1.4.12 | 192.168.1.3 | 973 | 0x80000003 | 0x002F84 |
| 10.1.4.12 | 192.168.1.4 | 1272 | 0x80000003 | 0x001F94 |
| 10.2.1.0 | 192.168.1.3 | 973 | 0x80000003 | 0x00CEEF |
| 10.2.1.0 | 192.168.1.4 | 1013 | 0x80000003 | 0x00C8F4 |
| 10.2.2.0 | 192.168.1.3 | 973 | 0x80000003 | 0x00C3F9 |
| 10.2.2.0 | 192.168.1.4 | 1013 | 0x80000003 | 0x00BDFE |
| 10.2.3.0 | 192.168.1.3 | 973 | 0x80000003 | 0x00B804 |
| 10.2.3.0 | 192.168.1.4 | 1013 | 0x80000003 | 0x00B209 |
| 10.2.4.0 | 192.168.1.3 | 973 | 0x80000003 | 0x00AD0E |
| 10.2.4.0 | 192.168.1.4 | 1013 | 0x80000003 | 0x00A713 |
| 10.2.5.0 | 192.168.1.3 | 973 | 0x80000003 | 0x00A218 |
| 10.2.5.0 | 192.168.1.4 | 1013 | 0x80000003 | 0x009C1D |
| 10.2.6.0 | 192.168.1.3 | 973 | 0x80000003 | 0x009722 |
| 10.2.6.0 | 192.168.1.4 | 1013 | 0x80000003 | 0x009127 |
| 10.2.7.0 | 192.168.1.3 | 973 | 0x80000003 | 0x008C2C |
| 10.2.7.0 | 192.168.1.4 | 1013 | 0x80000003 | 0x008631 |
| 10.2.8.0 | 192.168.1.3 | 973 | 0x80000003 | 0x008136 |
| 10.2.8.0 | 192.168.1.4 | 1013 | 0x80000003 | 0x007B3B |
| 10.2.9.0 | 192.168.1.3 | 973 | 0x80000003 | 0x007640 |
| 10.2.9.0 | 192.168.1.4 | 1013 | 0x80000003 | 0x007045 |
| 10.2.10.0 | 192.168.1.3 | 973 | 0x80000003 | 0x006B4A |
| 10.2.10.0 | 192.168.1.4 | 1013 | 0x80000003 | 0x00654F |

| 192.168.1.5 | 192.168.1.3 | 294 | 0x80000005 | 0x0081D8 |
| 192.168.1.5 | 192.168.1.4 | 1013 | 0x80000003 | 0x007FDB |
| 192.168.1.6 | 192.168.1.3 | 973 | 0x80000003 | 0x007BDF |
| 192.168.1.6 | 192.168.1.4 | 1013 | 0x80000003 | 0x0075E4 |

Router Link States (Area 1)

| Link ID | ADV Router | Age | Seq# | Checksum | Link count |
|---|---|---|---|---|---|
| 192.168.1.3 | 192.168.1.3 | 301 | 0x80000005 | 0x002CC8 | 2 |
| 192.168.1.4 | 192.168.1.4 | 1014 | 0x80000005 | 0x003DA5 | 2 |
| 192.168.1.5 | 192.168.1.5 | 301 | 0x80000007 | 0x009AB2 | 13 |
| 192.168.1.6 | 192.168.1.6 | 893 | 0x80000005 | 0x00EE33 | 13 |

Net Link States (Area 1)

| Link ID | ADV Router | Age | Seq# | Checksum |
|---|---|---|---|---|
| 10.1.4.1 | 192.168.1.3 | 301 | 0x80000001 | 0x00F6ED |
| 10.1.4.5 | 192.168.1.4 | 1014 | 0x80000003 | 0x00CE0E |
| 10.1.4.9 | 192.168.1.3 | 973 | 0x80000003 | 0x00B029 |
| 10.1.4.13 | 192.168.1.4 | 1014 | 0x80000003 | 0x008C47 |
| 10.2.1.2 | 192.168.1.5 | 841 | 0x80000003 | 0x0026B5 |
| 10.2.2.2 | 192.168.1.5 | 841 | 0x80000003 | 0x001BBF |
| 10.2.3.2 | 192.168.1.5 | 841 | 0x80000003 | 0x0010C9 |
| 10.2.4.2 | 192.168.1.5 | 841 | 0x80000003 | 0x0005D3 |
| 10.2.5.2 | 192.168.1.5 | 841 | 0x80000003 | 0x00F9DD |
| 10.2.6.2 | 192.168.1.5 | 841 | 0x80000003 | 0x00EEE7 |
| 10.2.7.2 | 192.168.1.5 | 841 | 0x80000003 | 0x00E3F1 |
| 10.2.8.2 | 192.168.1.5 | 841 | 0x80000003 | 0x00D8FB |
| 10.2.9.2 | 192.168.1.5 | 841 | 0x80000003 | 0x00CD06 |
| 10.2.10.2 | 192.168.1.5 | 841 | 0x80000003 | 0x00C210 |

Summary Net Link States (Area 1)

| Link ID | ADV Router | Age | Seq# | Checksum |
|---|---|---|---|---|
| 0.0.0.0 | 192.168.1.3 | 1227 | 0x80000003 | 0x008846 |
| 0.0.0.0 | 192.168.1.4 | 1273 | 0x80000003 | 0x00824B |

Type-5 AS External Link States

| Link ID | ADV Router | Age | Seq# | Checksum | Tag |
|---|---|---|---|---|---|
| 0.0.0.0 | 192.168.1.2 | 1240 | 0x80000003 | 0x00182C | 1 |
| 10.1.1.0 | 192.168.1.2 | 1240 | 0x80000003 | 0x001910 | 0 |
| 10.1.1.4 | 192.168.1.2 | 1240 | 0x80000003 | 0x00F034 | 0 |

10）查看交换机 sw5 的路由信息。

```
sw5#sho ip route
Codes: C - connected, S - static, R - RIP, M - mobile, B - BGP
       D - EIGRP, EX - EIGRP external, O - OSPF, IA - OSPF inter area
       N1 - OSPF NSSA external type 1, N2 - OSPF NSSA external type 2
       E1 - OSPF external type 1, E2 - OSPF external type 2
```

```
       i - IS-IS, su - IS-IS summary, L1 - IS-IS level-1, L2 - IS-IS level-2
       ia - IS-IS inter area, * - candidate default, U - per-user static route
       o - ODR, P - periodic downloaded static route

Gateway of last resort is 10.1.4.5 to network 0.0.0.0
     10.0.0.0/8 is variably subnetted, 14 subnets, 2 masks
C        10.2.8.0/24 is directly connected, Vlan8
C        10.2.9.0/24 is directly connected, Vlan9
C        10.2.10.0/24 is directly connected, Vlan10
O        10.1.4.12/30 [110/2] via 10.2.10.3, 00:05:47, Vlan10
                     [110/2] via 10.2.9.3, 00:05:47, Vlan9
                     [110/2] via 10.2.7.3, 00:05:47, Vlan7
                     [110/2] via 10.1.4.5, 00:05:47, FastEthernet1/5
O        10.1.4.8/30 [110/2] via 10.2.10.3, 00:05:47, Vlan10
                     [110/2] via 10.2.9.3, 00:05:47, Vlan9
                     [110/2] via 10.2.7.3, 00:05:47, Vlan7
                     [110/2] via 10.2.3.3, 00:05:47, Vlan3
C        10.2.1.0/24 is directly connected, Vlan1
C        10.2.2.0/24 is directly connected, Vlan2
C        10.2.3.0/24 is directly connected, Vlan3
C        10.1.4.4/30 is directly connected, FastEthernet1/5
C        10.2.4.0/24 is directly connected, Vlan4
C        10.2.5.0/24 is directly connected, Vlan5
C        10.2.6.0/24 is directly connected, Vlan6
C        10.2.7.0/24 is directly connected, Vlan7
C        10.1.4.0/30 is directly connected, FastEthernet1/4
     192.168.1.0/32 is subnetted, 2 subnets
C        192.168.1.5 is directly connected, Loopback0
O        192.168.1.6 [110/2] via 10.2.10.3, 00:05:48, Vlan10
                     [110/2] via 10.2.9.3, 00:05:48, Vlan9
                     [110/2] via 10.2.7.3, 00:05:48, Vlan7
                     [110/2] via 10.2.3.3, 00:05:48, Vlan3
O*IA 0.0.0.0/0 [110/2] via 10.1.4.5, 00:05:48, FastEthernet1/5
               [110/2] via 10.1.4.1, 00:05:48, FastEthernet1/4
```

11）查看交换机 sw5 的 OSPF 的邻居信息。

```
sw5# show ip ospf neighbor
Neighbor ID     Pri    State        Dead Time    Address      Interface
192.168.1.6     1      FULL/BDR     00:00:33     10.2.10.3    vlan10
192.168.1.6     1      FULL/BDR     00:00:33     10.2.9.3     vlan9
192.168.1.6     1      FULL/BDR     00:00:33     10.2.8.3     vlan8
192.168.1.6     1      FULL/BDR     00:00:33     10.2.7.3     vlan7
192.168.1.6     1      FULL/BDR     00:00:33     10.2.6.3     vlan6
192.168.1.6     1      FULL/BDR     00:00:33     10.2.5.3     vlan5
192.168.1.6     1      FULL/BDR     00:00:33     10.2.4.3     vlan4
192.168.1.6     1      FULL/BDR     00:00:33     10.2.3.3     vlan3
192.168.1.6     1      FULL/BDR     00:00:33     10.2.2.3     vlan2
192.168.1.6     1      FULL/BDR     00:00:33     10.2.1.3     vlan1
```

| | | | | | | |
|---|---|---|---|---|---|---|
| 192.168.1.4 | 100 | FULL/DR | 00:00:38 | 10.1.4.5 | FastEthernet1/5 |
| 192.168.1.3 | 100 | FULL/DR | 00:00:36 | 10.1.4.1 | FastEthernet1/4 |

12）查看交换机 sw5 的 OSPF 的数据库信息。

```
sw5#show ip ospf database
```

OSPF Router with ID (192.168.1.5) (Process ID 1)

Router Link States (Area 1)

| Link ID | ADV Router | Age | Seq# | Checksum | Link count |
|---|---|---|---|---|---|
| 192.168.1.3 | 192.168.1.3 | 451 | 0x80000005 | 0x002CC8 | 2 |
| 192.168.1.4 | 192.168.1.4 | 1166 | 0x80000005 | 0x003DA5 | 2 |
| 192.168.1.5 | 192.168.1.5 | 451 | 0x80000007 | 0x009AB2 | 13 |
| 192.168.1.6 | 192.168.1.6 | 1043 | 0x80000005 | 0x00EE33 | 13 |

Net Link States (Area 1)

| Link ID | ADV Router | Age | Seq# | Checksum |
|---|---|---|---|---|
| 10.1.4.1 | 192.168.1.3 | 451 | 0x80000001 | 0x00F6ED |
| 10.1.4.5 | 192.168.1.4 | 1166 | 0x80000003 | 0x00CE0E |
| 10.1.4.9 | 192.168.1.3 | 1124 | 0x80000003 | 0x00B029 |
| 10.1.4.13 | 192.168.1.4 | 1166 | 0x80000003 | 0x008C47 |
| 10.2.1.2 | 192.168.1.5 | 991 | 0x80000003 | 0x0026B5 |
| 10.2.2.2 | 192.168.1.5 | 991 | 0x80000003 | 0x001BBF |
| 10.2.3.2 | 192.168.1.5 | 991 | 0x80000003 | 0x0010C9 |
| 10.2.4.2 | 192.168.1.5 | 991 | 0x80000003 | 0x0005D3 |
| 10.2.5.2 | 192.168.1.5 | 991 | 0x80000003 | 0x00F9DD |
| 10.2.6.2 | 192.168.1.5 | 991 | 0x80000003 | 0x00EEE7 |
| 10.2.7.2 | 192.168.1.5 | 991 | 0x80000003 | 0x00E3F1 |
| 10.2.8.2 | 192.168.1.5 | 991 | 0x80000003 | 0x00D8FB |
| 10.2.9.2 | 192.168.1.5 | 991 | 0x80000003 | 0x00CD06 |
| 10.2.10.2 | 192.168.1.5 | 991 | 0x80000003 | 0x00C210 |

Summary Net Link States (Area 1)

| Link ID | ADV Router | Age | Seq# | Checksum |
|---|---|---|---|---|
| 0.0.0.0 | 192.168.1.3 | 1378 | 0x80000003 | 0x008846 |
| 0.0.0.0 | 192.168.1.4 | 1425 | 0x80000003 | 0x00824B |

13）查看交换机 sw5 的 VRRP 信息。

```
sw5#show vrrp brief
```

| Interface | Grp | Pri | Time | Own | Pre State | Master addr | Group addr |
|---|---|---|---|---|---|---|---|
| vl1 | 1 | 120 | 3531 | Y | Master | 10.2.1.2 | 10.2.1.1 |
| vl2 | 2 | 100 | 3609 | Y | Backup | 10.2.2.3 | 10.2.2.1 |
| vl3 | 3 | 120 | 3531 | Y | Master | 10.2.3.2 | 10.2.3.1 |
| vl4 | 4 | 100 | 3609 | Y | Backup | 10.2.4.3 | 10.2.4.1 |
| vl5 | 5 | 120 | 3531 | Y | Master | 10.2.5.2 | 10.2.5.1 |
| vl6 | 6 | 100 | 3609 | Y | Backup | 10.2.6.3 | 10.2.6.1 |
| vl7 | 7 | 120 | 3531 | Y | Master | 10.2.7.2 | 10.2.7.1 |

| | | | | | | | |
|---|---|---|---|---|---|---|---|
| vl8 | | 8 | 100 3609 | Y | Backup | 10.2.8.3 | 10.2.8.1 |
| vl9 | | 9 | 120 3531 | Y | Master | 10.2.9.2 | 10.2.9.1 |
| vl10 | | 10 | 100 3609 | Y | Backup | 10.2.10.3 | 10.2.10.1 |

14）查看交换机 sw6 的路由信息。

```
sw6#sho ip route
Codes: C - connected, S - static, R - RIP, M - mobile, B - BGP
       D - EIGRP, EX - EIGRP external, O - OSPF, IA - OSPF inter area
       N1 - OSPF NSSA external type 1, N2 - OSPF NSSA external type 2
       E1 - OSPF external type 1, E2 - OSPF external type 2
       i - IS-IS, su - IS-IS summary, L1 - IS-IS level-1, L2 - IS-IS level-2
       ia - IS-IS inter area, * - candidate default, U - per-user static route
       o - ODR, P - periodic downloaded static route

Gateway of last resort is 10.1.4.13 to network 0.0.0.0
     10.0.0.0/8 is variably subnetted, 14 subnets, 2 masks
C       10.2.8.0/24 is directly connected, Vlan8
C       10.2.9.0/24 is directly connected, Vlan9
C       10.2.10.0/24 is directly connected, Vlan10
C       10.1.4.12/30 is directly connected, FastEthernet1/4
C       10.1.4.8/30 is directly connected, FastEthernet1/5
C       10.2.1.0/24 is directly connected, Vlan1
C       10.2.2.0/24 is directly connected, Vlan2
C       10.2.3.0/24 is directly connected, Vlan3
O       10.1.4.4/30 [110/2] via 10.2.10.2, 00:08:01, Vlan10
                   [110/2] via 10.2.9.2, 00:08:01, Vlan9
                   [110/2] via 10.2.7.2, 00:08:01, Vlan7
                   [110/2] via 10.1.4.13, 00:08:01, FastEthernet1/4
C       10.2.4.0/24 is directly connected, Vlan4
C       10.2.5.0/24 is directly connected, Vlan5
C       10.2.6.0/24 is directly connected, Vlan6
C       10.2.7.0/24 is directly connected, Vlan7
O       10.1.4.0/30 [110/2] via 10.2.10.2, 00:08:01, Vlan10
                   [110/2] via 10.2.9.2, 00:08:01, Vlan9
                   [110/2] via 10.2.7.2, 00:08:01, Vlan7
                   [110/2] via 10.2.3.2, 00:08:01, Vlan3
     192.168.1.0/32 is subnetted, 2 subnets
O       192.168.1.5 [110/2] via 10.2.10.2, 00:08:01, Vlan10
                   [110/2] via 10.2.9.2, 00:08:01, Vlan9
                   [110/2] via 10.2.7.2, 00:08:01, Vlan7
                   [110/2] via 10.2.3.2, 00:08:01, Vlan3
C       192.168.1.6 is directly connected, Loopback0
O*IA 0.0.0.0/0 [110/2] via 10.1.4.13, 00:08:01, FastEthernet1/4
               [110/2] via 10.1.4.9, 00:08:01, FastEthernet1/5
```

15）查看交换机 sw6 的 OSPF 的邻居信息。

```
sw6#show ip ospf neighbor
Neighbor ID     Pri   State       Dead Time   Address    Interface
192.168.1.5     1     FULL/DR     00:00:31    10.2.10.2  vlan10
192.168.1.5     1     FULL/DR     00:00:31    10.2.9.2   vlan9
192.168.1.5     1     FULL/DR     00:00:31    10.2.8.2   vlan8
192.168.1.5     1     FULL/DR     00:00:31    10.2.7.2   vlan7
192.168.1.5     1     FULL/DR     00:00:31    10.2.6.2   vlan6
```

| | | | | | | |
|---|---|---|---|---|---|---|
| 192.168.1.5 | 1 | FULL/DR | 00:00:31 | 10.2.5.2 | vlan5 |
| 192.168.1.5 | 1 | FULL/DR | 00:00:31 | 10.2.4.2 | vlan4 |
| 192.168.1.5 | 1 | FULL/DR | 00:00:31 | 10.2.3.2 | vlan3 |
| 192.168.1.5 | 1 | FULL/DR | 00:00:31 | 10.2.2.2 | vlan2 |
| 192.168.1.5 | 1 | FULL/DR | 00:00:31 | 10.2.1.2 | vlan1 |
| 192.168.1.3 | 100 | FULL/DR | 00:00:33 | 10.1.4.9 | FastEthernet1/5 |
| 192.168.1.4 | 100 | FULL/DR | 00:00:35 | 10.1.4.13 | FastEthernet1/4 |

16）查看交换机 sw6 的 OSPF 的数据库信息。

sw6#show ip ospf database

OSPF Router with ID (192.168.1.6) (Process ID 1)

Router Link States (Area 1)

| Link ID | ADV Router | Age | Seq# | Checksum | Link count |
|---|---|---|---|---|---|
| 192.168.1.3 | 192.168.1.3 | 548 | 0x80000005 | 0x002CC8 | 2 |
| 192.168.1.4 | 192.168.1.4 | 1262 | 0x80000005 | 0x003DA5 | 2 |
| 192.168.1.5 | 192.168.1.5 | 548 | 0x80000007 | 0x009AB2 | 13 |
| 192.168.1.6 | 192.168.1.6 | 1139 | 0x80000005 | 0x00EE33 | 13 |

Net Link States (Area 1)

| Link ID | ADV Router | Age | Seq# | Checksum |
|---|---|---|---|---|
| 10.1.4.1 | 192.168.1.3 | 547 | 0x80000001 | 0x00F6ED |
| 10.1.4.5 | 192.168.1.4 | 1262 | 0x80000003 | 0x00CE0E |
| 10.1.4.9 | 192.168.1.3 | 1220 | 0x80000003 | 0x00B029 |
| 10.1.4.13 | 192.168.1.4 | 1262 | 0x80000003 | 0x008C47 |
| 10.2.1.2 | 192.168.1.5 | 1088 | 0x80000003 | 0x0026B5 |
| 10.2.2.2 | 192.168.1.5 | 1088 | 0x80000003 | 0x001BBF |
| 10.2.3.2 | 192.168.1.5 | 1088 | 0x80000003 | 0x0010C9 |
| 10.2.4.2 | 192.168.1.5 | 1088 | 0x80000003 | 0x0005D3 |
| 10.2.5.2 | 192.168.1.5 | 1088 | 0x80000003 | 0x00F9DD |
| 10.2.6.2 | 192.168.1.5 | 1088 | 0x80000003 | 0x00EEE7 |
| 10.2.7.2 | 192.168.1.5 | 1088 | 0x80000003 | 0x00E3F1 |
| 10.2.8.2 | 192.168.1.5 | 1088 | 0x80000003 | 0x00D8FB |
| 10.2.9.2 | 192.168.1.5 | 1088 | 0x80000003 | 0x00CD06 |
| 10.2.10.2 | 192.168.1.5 | 1088 | 0x80000003 | 0x00C210 |

Summary Net Link States (Area 1)

| Link ID | ADV Router | Age | Seq# | Checksum |
|---|---|---|---|---|
| 0.0.0.0 | 192.168.1.3 | 1473 | 0x80000003 | 0x008846 |
| 0.0.0.0 | 192.168.1.4 | 1521 | 0x80000003 | 0x00824B |

17）查看交换机 sw6 的 VRRP 信息。

sw6# show vrrp brief

| Interface | Grp | Pri | Time | Own Pre | State | Master addr | Group addr |
|---|---|---|---|---|---|---|---|
| Vl1 | 1 | 100 | 3609 | Y | Backup | 10.2.1.2 | 10.2.1.1 |
| Vl2 | 2 | 120 | 3531 | Y | Master | 10.2.2.3 | 10.2.2.1 |
| Vl3 | 3 | 100 | 3609 | Y | Backup | 10.2.3.2 | 10.2.3.1 |
| Vl4 | 4 | 120 | 3531 | Y | Master | 10.2.4.3 | 10.2.4.1 |
| Vl5 | 5 | 100 | 3609 | Y | Backup | 10.2.5.2 | 10.2.5.1 |
| Vl6 | 6 | 120 | 3531 | Y | Master | 10.2.6.3 | 10.2.6.1 |

| | | | | | | | |
|---|---|---|---|---|---|---|---|
| Vl7 | 7 | 100 | 3609 | Y | Backup | 10.2.7.2 | 10.2.7.1 |
| Vl8 | 8 | 120 | 3531 | Y | Master | 10.2.8.3 | 10.2.8.1 |
| Vl9 | 9 | 100 | 3609 | Y | Backup | 10.2.9.2 | 10.2.9.1 |
| Vl10 | 10 | 120 | 3531 | Y | Master | 10.2.10.3 | 10.2.10.1 |

18）查看交换机 sw7 的 VLAN 信息。

```
sw7#show vlan-switch
```

| VLAN | Name | Status | Ports |
|---|---|---|---|
| 1 | default | active | Fa1/4, Fa1/5, Fa1/6, Fa1/7 |
| | | | Fa1/8, Fa1/9, Fa1/10, Fa1/11 |
| | | | Fa1/12, Fa1/13, Fa1/14, Fa1/15 |
| 2 | VLAN0002 | active | Fa1/0 |
| 3 | VLAN0003 | active | Fa1/1 |
| 4 | VLAN0004 | active | |
| 5 | VLAN0005 | active | |
| 6 | VLAN0006 | active | |
| 7 | VLAN0007 | active | |
| 8 | VLAN0008 | active | |
| 9 | VLAN0009 | active | |
| 10 | VLAN0010 | active | |
| 1002 | fddi-default | active | |
| 1003 | token-ring-default | active | |
| 1004 | fddinet-default | active | |
| 1005 | trnet-default | active | |

| VLAN | Type | SAID | MTU | Parent | RingNo | BridgeNo | Stp | BrdgMode | Trans1 | Trans2 |
|---|---|---|---|---|---|---|---|---|---|---|
| 1 | enet | 100001 | 1500 | - | - | - | - | - | 1002 | 1003 |

| VLAN | Type | SAID | MTU | Parent | RingNo | BridgeNo | Stp | BrdgMode | Trans1 | Trans2 |
|---|---|---|---|---|---|---|---|---|---|---|
| 2 | enet | 100002 | 1500 | - | - | - | - | - | 0 | 0 |
| 3 | enet | 100003 | 1500 | - | - | - | - | - | 0 | 0 |
| 4 | enet | 100004 | 1500 | - | - | - | - | - | 0 | 0 |
| 5 | enet | 100005 | 1500 | - | - | - | - | - | 0 | 0 |
| 6 | enet | 100006 | 1500 | - | - | - | - | - | 0 | 0 |
| 7 | enet | 100007 | 1500 | - | - | - | - | - | 0 | 0 |
| 8 | enet | 100008 | 1500 | - | - | - | - | - | 0 | 0 |
| 9 | enet | 100009 | 1500 | - | - | - | - | - | 0 | 0 |
| 10 | enet | 100010 | 1500 | - | - | - | - | - | 0 | 0 |
| 1002 | fddi | 101002 | 1500 | - | - | - | - | - | 1 | 1003 |
| 1003 | tr | 101003 | 1500 | 1005 | 0 | - | - | srb | 1 | 1002 |
| 1004 | fdnet | 101004 | 1500 | - | - | 1 | ibm | - | 0 | 0 |
| 1005 | trnet | 101005 | 1500 | - | - | 1 | ibm | - | 0 | 0 |

19）查看交换机 sw8 的 VLAN 信息。

```
sw8#show vlan-switch
```

| VLAN | Name | Status | Ports |
|---|---|---|---|
| 1 | default | active | Fa1/4, Fa1/5, Fa1/6, Fa1/7 |

|   |   |   |   |   |
|---|---|---|---|---|
|   |   |   |   | Fa1/8, Fa1/9, Fa1/10, Fa1/11 |
|   |   |   |   | Fa1/12, Fa1/13, Fa1/14, Fa1/15 |
| 2 | VLAN0002 |   | active | Fa1/0 |
| 3 | VLAN0003 |   | active | Fa1/1 |
| 4 | VLAN0004 |   | active |   |
| 5 | VLAN0005 |   | active |   |
| 6 | VLAN0006 |   | active |   |
| 7 | VLAN0007 |   | active |   |
| 8 | VLAN0008 |   | active |   |
| 9 | VLAN0009 |   | active |   |
| 10 | VLAN0010 |   | active |   |
| 1002 | fddi-default |   | active |   |
| 1003 | token-ring-default |   | active |   |
| 1004 | fddinet-default |   | active |   |
| 1005 | trnet-default |   | active |   |

| VLAN | Type | SAID | MTU | Parent | RingNo | BridgeNo | Stp | BrdgMode | Trans1 | Trans2 |
|------|------|------|-----|--------|--------|----------|-----|----------|--------|--------|
| 1 | enet | 100001 | 1500 | - | - | - | - | - | 1002 | 1003 |

| VLAN | Type | SAID | MTU | Parent | RingNo | BridgeNo | Stp | BrdgMode | Trans1 | Trans2 |
|------|------|------|-----|--------|--------|----------|-----|----------|--------|--------|
| 2 | enet | 100002 | 1500 | - | - | - | - | - | 0 | 0 |
| 3 | enet | 100003 | 1500 | - | - | - | - | - | 0 | 0 |
| 4 | enet | 100004 | 1500 | - | - | - | - | - | 0 | 0 |
| 5 | enet | 100005 | 1500 | - | - | - | - | - | 0 | 0 |
| 6 | enet | 100006 | 1500 | - | - | - | - | - | 0 | 0 |
| 7 | enet | 100007 | 1500 | - | - | - | - | - | 0 | 0 |
| 8 | enet | 100008 | 1500 | - | - | - | - | - | 0 | 0 |
| 9 | enet | 100009 | 1500 | - | - | - | - | - | 0 | 0 |
| 10 | enet | 100010 | 1500 | - | - | - | - | - | 0 | 0 |
| 1002 | fddi | 101002 | 1500 | - | - | - | - | - | 1 | 1003 |
| 1003 | tr | 101003 | 1500 | 1005 | 0 | - | - | srb | 1 | 1002 |
| 1004 | fdnet | 101004 | 1500 | - | - | 1 | ibm | - | 0 | 0 |
| 1005 | trnet | 101005 | 1500 | - | - | 1 | ibm | - | 0 | 0 |

# 附录　Dynamips 模拟器的使用

## 一、介绍

Dynamips 是 Christophe Fillot 编写的一款 Cisco 路由器模拟器。它模拟 1700、2600、3600、3700 和 7200 硬件平台，并且运行标准的 IOS 映像文件。

Dynamips 模拟器的用途：

（1）作为培训的平台使用，通过软件的方式模拟使用真实环境中的设备。它可以让大家更熟悉 Cisco 的设备，因为 Cisco 是网络技术的领导者。

（2）测试和实验 Cisco IOS 的各种特性。

（3）迅速检测实施到真实路由上的配置。

Dynamips 提供了通过加载 NM-16ESW 模块模拟简单的虚拟交换机，但是不能模拟 Catalyst 交换机。NM-16ESW 是一块 16 口的网络模块，具备交换功能。有很多交换机的特性在这个卡上是不能实现的。

Dynagen 是 Dynamips 的一个基于文本的前端控制系统，它采用 Hypervisor 超级监控模式和 Dynamips 通信，Dynagen 的作用如下：

（1）使用简单，特定的虚拟路由器硬件的配置文件也容易理解。

（2）路由器、网桥、帧中继、ATM 和以太网交换机互连的语法简单。

（3）可以工作在客户端/服务器（C/S）模式下，可以让运行在工作站上的 Dynagen 和运行在后台服务器上的 Dynamips 通信。Dynagen 也可以同时使用多个分布式的 Dynamips 服务器来运行一个大的虚拟网络，当然可以运行在同一个系统上。

（4）提供了 CLI 的管理方式，可以列出设备、启动、停止、重启、挂起、恢复和连接虚拟路由器的 console 口。

## 二、安装

首先，安装 Libpcap 或者 Winpcap，这取决于你将要运行 Dynamips 的机器的平台。这个库用来提供桥接路由器接口到物理网卡，就是将物理网卡作为虚拟路由器的网卡，可以让虚拟路由器和真实环境中的设备通信。Windows 用户必须安装 Winpcap4.0 或者更新的版本，Linux 用户必须安装相应的 Libpcap 版本。

然后，Windows 用户应该安装 Dynagen-0.11.0_win_setup.exe 安装包。在本地或远程设备上运行 Dynamips/Dynagen 所需要的。安装步骤如下：将鼠标指针移动到 Dynagen-0.11.0_win_setup.exe 安装包上，双击鼠标左键，弹出如图 f-1 所示的对话框。

单击 Next 按钮，弹出如图 f-2 所示的对话框。

单击 I Agree 按钮，弹出如图 f-3 所示的对话框。

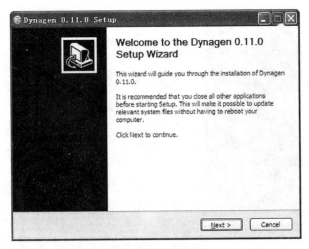

图 f-1　安装步骤 1

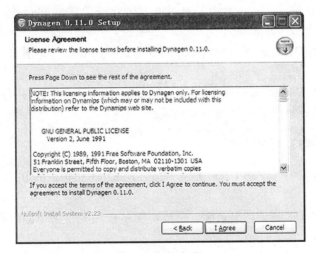

图 f-2　安装步骤 2

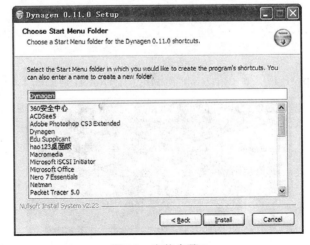

图 f-3　安装步骤 3

单击 Install 按钮，弹出如图 f-4 所示的对话框。

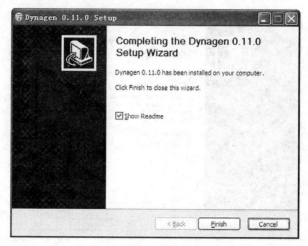

图 f-4　安装步骤 4

单击 Finish 按钮，完成 Dynagen-0.11.0_win_setup.exe 安装包的安装。

Linux 用户应该下载 Dynamips/Dynagen 压缩包，并解压到一个合适的目录中。然后在 /usr/local/bin 中创建到 Dynagen 和 Dynamips 可执行的 Symlinks，或者是在其他的 PATH 中做这件事。

**注意**：建议大家在 Windows 的命令行窗口中用输入命令的方式来执行。Dynagen/Dynamips 是绿色软件，可以 copy 到任何一个地方直接运行。

### 三、硬件平台的模拟

Dynamips 运行时需要真实的 Cisco 1700、2691、3620、3640、3660、3725、3745 或者 7200 的 IOS 软件。

Dynamips 究竟支持哪些硬件平台呢？又支持哪些模块呢？用下面的命令体现：

dynamips -P 1700

dynamips -P 2600

dynamips -P 3600

dynamips -P 3725

dynamips -P 3745

dynamips -P 7200

例如输入 dynamips -P 1700，在后面的输出信息中你会看到如图 f-5 所示的信息。

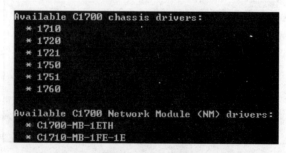

图 f-5　平台信息

说明 Dynamips 对 1700 系列的路由器可以支持的有：1710、1720、1721、1750、1751、

1760。支持的模块有 C1700-MB-1ETH 和 C1710-MB-1FE-1E。

3700 的 slot0 只能安装 GT96100-FE，7200 只能安装 C7200-IO-FE。在 3600 系列路由器的 slot0 里面可以安装 NM -XX 的卡。

### 四、网络文件

使用 Dynagen 一定离不开 .net 文件——网络文件。.net 文件决定了所有 Dynamips 的参数，包括 Dynamips 模拟出来的路由器的拓扑结构。用文本编辑器可以打开 simple1.net 文件。

任何以 "#" 开始的都是注释行，并会被忽略。

下面来看 s1.net 文件：
=====================================
```
# Simple lab
autostart = false    #启动网络文件时，所有模拟的设备状态为 stop
ghostios   = true
sparsemem = true     #此行参数和上一行的 ghostios 参数起的作用是节省内存资源

[localhost:7200]   #设置运行 Dynamips 服务器的 IP 地址或域名和端口，localhost 为本机，默认端口为 7200

    [[3745]]   #设置模拟器模拟的设备型号
    image = ..\c3745-adventerprisek9-mz.124-16.bin   #设置加载模拟设备的镜像文件
    ram = 160   #设置模拟设备的内存大小

    [[ROUTER sw1]]   #创建一台模拟路由器，名称为 sw1
    model = 3745   #设置创建模拟设备的型号
    slot1 = NM-16ESW   #在 slot1 插槽加载 NM-16ESW 模块
    f1/0 = sw2 f1/0    #将 sw1 的 f1/0 端口与 sw2 的 f1/0 端口连接起来
    f1/1 = sw2 f1/1
    f1/2 = sw3 f1/2
    f1/3 = sw3 f1/3

    [[ROUTER sw2]]
    model = 3745
    slot1 = NM-16ESW
    f1/2 = sw3 f1/0
    f1/3 = sw3 f1/1

    [[ROUTER sw3]]
    model = 3745
    slot1 = NM-16ESW

    [[ROUTER sw4]]
    model = 3745
    slot1 = NM-16ESW
    f1/0 = sw1 f1/4
    f1/1 = sw1 f1/5
```
=====================================

[localhost]指定运行 Dynamips 的主机是 localhost。在这个例子中，我们要在同一台机器上

运行 Dynamips 和 Dynagen，因此，我们指定 localhost。如果在不同的机器上运行 Dynamips，则应该写上：在这里用机器的 hostname 或者 IP 地址来替代 localhost。由于 Windows 操作系统对单一进程使用的逻辑内存有一定的限制。如果运行过多的路由器，会导致无法启动。有两种方式可以解决这个问题：

（1）使用分布式环境，在多台 PC 上运行多个 Dynamips。

（2）在一台主机上启用多个 Dynamips。每个 Dynamips 下面运行一定数量的路由器。最关键的就是在[localhost]中指定使用哪个 Dynamips 服务器。可以这样写：

[localhost:7200]
[localhost:7201]
[localhost:7202]

默认就是 7200。这个 7200 和 dynamips－H 7200 中的 7200 是一致的。看下面的例子：

[localhost:7200]
    udp = 10000

[[7200]]
    image = ios.bin
    exec_area = 32
    npe = npe-400
    ram = 160

[[router R1]]
    s1/0 = R2 s1/0

[localhost:7201]
    udp = 20000

[[7200]]
    image = ios.bin
    exec_area = 32
    npe = npe-400
    ram = 160

[[router R2]]
========================================

### 五、运行虚拟实验

为了运行这个虚拟实验，首先在本地机器上启动 Dynamips 服务器。在 Windows 命令行下，进入 Dynamips 的所在目录 C:\Progam Files\Dynamips，使用命令 dynamips－H 7200 启动 Dynamips Server；也可以双击鼠标运行 dynamips-start.cmd 文件。默认情况下，.net 网络文件中[localhost]默认侦听的端口是 7200。因此一般情况下我们启动 Dynamips 的时候，H 后面的参数为 7200，如图 f-6 所示。

在 Linux/UNIX 系统上，在后台启动服务器。例如：

nice dynamips –H 7200 &

在 Windows 系统上，打开一个.net 网络文件时最简单的办法就是双击.net 文件，然后选择 dg-local 程序启动，如图 f-7 所示。

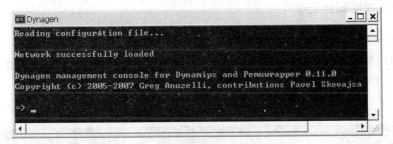

图 f-6  Dynamips 服务器

图 f-7  Dynagen 控制台

在 Linux/UNIX 系统中，可以使用命令来启动虚拟实验：

dynagen simple1.net

可能会看到关于 no idle-pc value 的告警信息，现在先将其忽略，后面有关于 idle-pc 的详细说明。查看所有虚拟实验中的设备，使用 list 命令，如图 f-8 所示。

图 f-8  查看控制台中设备

上图说明有 4 台路由器：sw1、sw2、sw3 和 sw4。目前它们都还没有启动。sw1 的 console 口是 TCP 端口 2000，sw2 的 console 口是 TCP 端口 2001。下面输入 start sw1 命令就可以启动路由器 sw1，如图 f-9 所示。

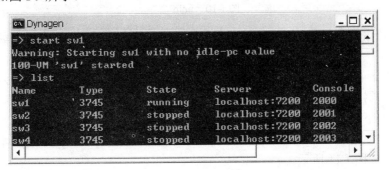

图 f-9  启动设备 sw1

通过 Telnet 命令就能登录到虚拟路由器上；或者如果配置 dynagen.ini 文件来指定 Telnet 客户端，只要输入 telnet sw1，就可以连接到 sw1 的 console 口（注：命令和设备名称是区分大小写的）。

甚至更方便的是，你可以输入 console /all，每一个虚拟路由器都将有一个连接 console 的窗口。如果使用 Linux、OS X 或者 Windows 的 Tera Term SSH，console /all 会工作得更好，因为标题栏有路由器的名字。然而 Windows 内置的 Telnet 命令不允许这个。因此用最初的 console /all 连接实验中的路由器，使用 Windows 内置的 telnet 命令可能不是一个好方法。因为可能无法知道你连接的是哪个路由器。telnet /all 命令和 console /all 的效果一样。

### 六、通过管理控制台工作

1. help 命令介绍

通过管理控制台，使用 help 命令或者 '？' 可以查看所有可以使用的命令，如图 f-10 所示。

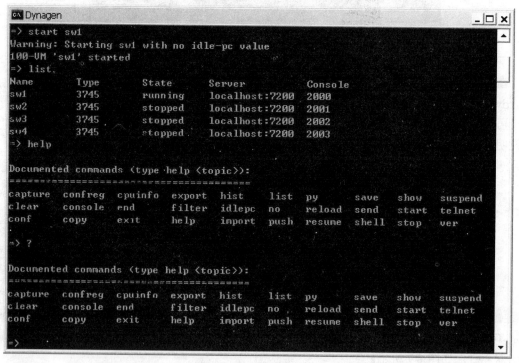

图 f-10　help 命令

获得一个特定命令的帮助，可以输入 help command 或者 command ？。例如：help console 或 console？。

2. stop 命令介绍

对虚拟路由器"断电"可使用 stop 命令。

命令格式如下：stop {/all | router1 [router2] ...}

关闭一个单独的路由器使用 stop routername 命令；也可以让所有路由器停止工作，使用 stop /all 命令，如图 f-11 所示。

图 f-11 关闭设备命令

3. start 命令介绍

启动虚拟路由器用 start 命令。

命令格式如下：start {/all | router1 [router2] ...}

4. reload 命令介绍

重启虚拟路由器用 reload 命令，如图 f-12 所示。

命令格式如下：reload{/all | router1 [router2] ...}

图 f-12 重启设备命令

5. suspend 和 resume 命令介绍

临时暂停指定路由器的命令为 suspend，启动临时暂停指定路由器的命令为 resume。suspend 与 resume 命令语法与 stop 和 start 相似，如图 f-13 所示。

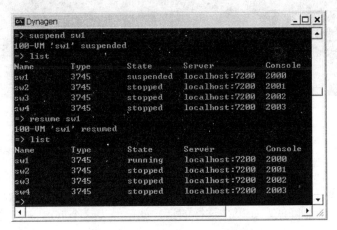

图 f-13 suspend 和 resume 命令

#### 6. exit 命令介绍

exit 命令用于从网络上停止并删除所有的设备，并且退出 Dynagen。

### 七、计算 Idle-PC 值

前面的实验会导致系统的 CPU 高居 100%。Idlepc 可以大幅度地降低主机的 CPU 消耗。如何计算 Idlepc 的值呢？首先，打开一个实验，确认一台路由器进程正在运行；然后，telnet 到正在运行的路由器进程。如果发现 IOS 自动配置的提示，用 no 来回复，如图 f-14 所示。

图 f-14 设备自动配置

等所有的接口都初始化以后，再等一会儿，确保路由器不再运行并处于空闲状态。这时路由器应如图 f-15 所示那样。

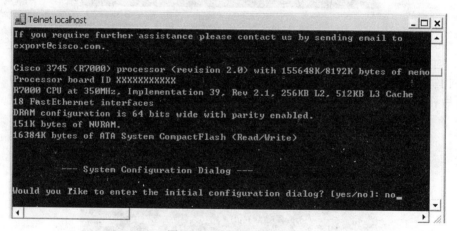

图 f-15 设备启动结束

现在，切换到 Dynagen 的管理控制台，使用 idlepc get routername 命令，会看到一条关于收集统计信息的消息，大概 10 秒钟以后，会看到 Idlepc 值的一个统计列表。

带星号 "*" 的值是更好一些的，选择菜单中的一个序号值并按 Enter 键。此时可以看到主机的 CPU 利用率明显降低。如果是这样的话，就已经为这个 IOS 映象文件获得了一个良好的 Idlepc 值。

如果 CPU 利用率没有降低，请试一下另外的 Idlepc 值。输入 idlepc show routername 命令显示刚才的 Idlepc 值，这次请选一个与上次不同的值。

新的 Idlepc 值马上生效。一旦发现一个不错的值，可以保存到本地 Idlepc 数据中，使用命令 idlepc save routername db。通过 dynagen.ini 中的 idledb 选项保存 IOS 映象文件的 Idlepc 值。默认的文件名是 dynagenidledb.ini，在$HOME 或者 documents and settings 目录下面可以找到这个文件。

一旦 Idlepc 值存在于数据中，在任何实验中，只要使用这个 IOS 映像文件，它都会被自动应用。如果 Dynagen 启动一台路由器的时候没有发现 Idlepc 值，它会给出 Warning: Starting xxx with no idle-pc value 的告警信息。如果你愿意在网络文件中存储 Idlepc 值，使用 idlepc save routername 命令获得 Idlepc 值，然后在网络文件中添加一行"idlepc = xxxx"到路由器定义小节中（如[[ROUTER R1]]）；或者使用 idlepc save routername default 保存到网络文件的路由器平台默认小节中。

### 八、与真实网络通信

Dynamips 可以桥接虚拟路由器接口到真实的主机接口，允许虚拟网络和真实世界通信。在 Linux 系统上，这被描述为 NIO_linux_eth。例如：

f0/0 = NIO_linux_eth:eth0

这个桥接路由器的 f0/0 接口连接到主机的 eth0 接口。数据包离开 f0/0 时通告 eth0 被发送到真实网络中，而且返回数据包同样也转发回虚拟路由器。

在 Windows 系统上，Winpcap 库用来完成这个桥接功能。在 Windows 系统上，接口规格要稍微复杂一些。因此在 Windows 主机上，Dynamips 提供一个命令行开关列出可用接口。Dynamips/Dynagen 的 Windows 安装包包括了一个这样的快捷方式。在桌面上，打开 Network Device List 快捷方式，或者使用下面的命令：

Dynamips -e

因此在 Windows 系统上，我们使用 F0/0 = NIO_gen_eth:\Device\NPF_{E2B31736-C716-4215-A9CB-ADA78C760F52}来桥接到本地以太网卡（不同网卡的{}中的值不同）。

### 九、C/S 和多服务器操作

Dynamips 的 Hyperviso 超级监控模式被 Dynagen 用来作为一个 TCP/IP 通信的通道，因此 Dynagen 客户端可以和 Dynamips 模拟器运行在不同的设备上。这个通过在网络文件中指定一个不同于 localhost 的主机来实现。

在 multiserver.net 的实验中，首先我们指定设备来运行本地系统（一个 Windows XP 主机）：

# A windows server (the local machine)
[xplt]
  [[7200]]
    image = \PROGRA~1\Dynamips\images\c7200-ik9o3s-mz.122-15.T17.image
    ram = 96
  [[ROUTER R1]]
  # Connect to s1/0 on R2 running on a different server
    s1/0 = R2 s1/0

注意：首先，必须使用 DNS 或者本地主机的 IP 地址，来区别系统而不是 localhost。这是因为下面定义的其他服务器将使用名字和本地系统通信；其次，连接到一个其他系统上的设备的方法和连接本地系统一样简单，可以使用任何 Dynamips 支持的连接方法或设备（以太网、串口、ATM、桥接、以太网交换机、FR 交换机等）。

接下来定义其他 Dynamips 服务器，而且路由器例程运行在服务器上：

```
# A linux server
[bender:7200]
workingdir = /home/greg/labs/dist1
[[7200]]
image = /opt/7200-images/c7200-ik9o3s-mz.122-15.T17.image
ram = 96
[[ROUTER R2]]
```

这里我们将与名为 bender 的服务器通信（也可以在这里指定 IP 地址，优于一个 DNS 名字）。我们可以指定一个 Dynamips 进程侦听的 TCP 端口——7200。

和远程服务器通信时，需要为实验指定工作目录。就像在前面实验中需要注意的一样，Dynamips 在工作目录中存储几个文件，包括虚拟路由器的 NVRAM、bootflash、logfiles 和其他一些工作文件。当在同一台机器上运行 Dynamips 和 Dynagen 时，不必为工作文件指定工作目录。但是在分布式网络环境中，网络文件在客户端，工作文件在主机上。因此需要为 Dynamips 主机指定一个完整的工作目录，确认在平台上使用了正确的目录分割符（这里 "/" 用于 Linux 系统）。

**注意**：在任何有防火墙的主机上运行 Dynamips 服务器（例如 XP SP2 的防火墙）要允许特定的流量，包括 Dynamips 服务器端口（默认是 TCP 7200）、console 端口（例如 TCP 2000、2001……）和 NIO 连接接口之间的端口，开始于 UDP 10000 并逐步增加。

当然，可以在很多台主机上启动 Dynamips，每台主机运行一定数量的路由器，做一个大的分布式网络。需要注意的是：

（1）每台主机上启动 Dynamips 的时候，侦听端口尽量不同。

（2）每台主机上 NIO 接口的 UPD 尽量不同。比如 Server1 上用 10000，Server2 上用 12000 等。

例如：

在运行 mul-eserver.net 文件之前，在 IP 地址为 192.168.7.5 的主机上运行：

```
c:\Program Files\Dynamips\dynamips –H 7200
```

在 IP 地址为 192.168.7.6 的主机上运行：

```
c:\Program Files\Dynamips\dynamips –H 7201
```

mul-eserver.net 文件可以在任何一台能够与 IP 地址为 192.168.7.5 和 192.168.7.6 且 TCP 端口分别为 7200 和 7201 的主机进行通信的主机上运行。

mul-eserver.net 内容为

```
# mul-eserver.net 文件开始

autostart = false
ghostios = true
sparsemem = true


[192.168.7.5:7200]

    [[7200]]
```

image = c:\Program Files\Dynamips\images\c7200-js-mz.123-5.bin
npe = npe-400
ram = 160
idlepc = 0x607204dc

[[ROUTER R1]]
slot1 = PA-4T+
s1/0 = F1 1

[[ROUTER R2]]
slot1 = PA-4T+
s1/0 = F1 2

[[ROUTER R3]]
slot1 = PA-4T+
s1/0 = F1 3

[[FRSW F1]]   #帧中继交换机
1:102 = 2:201     #冒号前面的数字是端口号，冒号后面的数字是 DLCI
1:103 = 3:301
2:203 = 3:302

[192.168.7.6:7201]
workingdir = e:\mut_server
　　[[3640]]
　　image =   c:\Program Files\Dynamips\images\c3640-ik9o3s-mz.123-26.bin
　　ram = 160

　　[[3660]]
　　image = ..\..\images\c3660-is-mz.124-13.bin
　　ram = 160

　　[[ROUTER R1]]
　　model = 3660
　　idlepc = 0x60611e60
　　f0/0 = switch3 f0/0
　　f0/1 = R1 F0/1

　　[[ROUTER switch1]]
　　model = 3640
　　idlepc = 0x60431654
　　slot0 = NM-16ESW

　　[[ROUTER switch3]]

        model = 3660
        idlepc = 0x60611e60
        slot1 = NM-16ESW
        f1/0 = switch1 f0/0
# mul-eserver.net 文件结束

### 十、数据包捕获

Dynamips/Dynagen 可以捕获虚拟的以太网卡或者串口的数据包，而且输出一个可以用 tcpdump、Wireshark 或者其他可以读取 Libpcap 捕获的文件格式的文件。

考虑一下三台路由器的情况，r1 和 r2 通过以太网互连，r2 和 r3 通过封装 HDLC 的 PPP 串口互连。网络文件如下：

model = 3660
[localhost]
[[3660]]
    image = c:\Program Files\Dynamips\images\c3660-ik9o3s-mz.124-10.image
    ram = 160

[[router r1]]
    f0/0 = r2 f0/0

[[router r2]]
    s1/0 = r3 s1/0

[[router r3]]

捕获 r1 的 f0/0 接口的流量，并输出到文件 r1.cap 中，在 Dynagen 窗口中输入下面的命令：
capture r1 f0/0 r1.cap
查看实时流量，用 Wireshark 打开 r1.cap 文件，如图 f-16 所示。

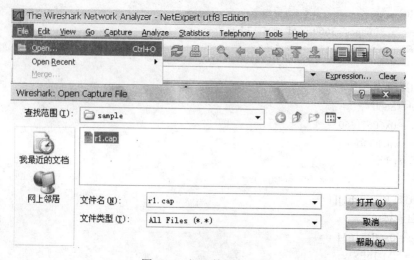

图 f-16　打开数据包文件

捕获会一直往输出文件中写入数据包。如果我们从 r1 ping r2，然后单击 reload this capture

file 图标，结果如图 f-17 所示。

| No. | Time | Source | Destination | Protocol | Info |
|---|---|---|---|---|---|
| 5 | 14.946000 | 192.168.1.1 | 192.168.1.2 | ICMP | Echo (ping) request |
| 6 | 14.985000 | 192.168.1.2 | 192.168.1.1 | ICMP | Echo (ping) reply |
| 7 | 15.024000 | 192.168.1.1 | 192.168.1.2 | ICMP | Echo (ping) request |
| 8 | 15.032000 | 192.168.1.2 | 192.168.1.1 | ICMP | Echo (ping) reply |
| 9 | 15.036000 | 192.168.1.1 | 192.168.1.2 | ICMP | Echo (ping) request |
| 10 | 15.039000 | 192.168.1.2 | 192.168.1.1 | ICMP | Echo (ping) reply |
| 11 | 15.043000 | 192.168.1.1 | 192.168.1.2 | ICMP | Echo (ping) request |
| 12 | 15.051000 | 192.168.1.2 | 192.168.1.1 | ICMP | Echo (ping) reply |
| 13 | 15.055000 | 192.168.1.1 | 192.168.1.2 | ICMP | Echo (ping) request |
| 14 | 15.059000 | 192.168.1.2 | 192.168.1.1 | ICMP | Echo (ping) reply |

图 f-17　查看 ICMP 数据包

停止捕获数据包，输入以下命令：

no capture r1 f0/0

Dynamips / Dynagen 也可以捕获串口的数据包。在下面这个案例中，我们还需要指定在路由器接口上的封装协议，这样 Wireshark 就会知道如何解码这些数据包。封装类型可以是 Frame-relay、HDLC 或者 PPP。要捕获 r2 和 r2 互连链路上的 HDLC 协议流量，执行以下命令：

capture r2 s1/0 r2.cap HDLC

通过 no capture r2 s1/0 命令来终止捕获。

**注意**：可以在多个路由器上同时捕获多个接口的数据。

# 参考文献

[1] http://www.ietf.org.
[2] http://www.ieee.org/.
[3] （美）史蒂文斯著．TCP/IP 详解．范建华译．北京：机械工业出版社，2000．
[4] （美）Matthew J.Castelli 著．网络工程师手册．袁国忠译．北京：人民邮电出版社，2005．
[5] （美）David Hucaby, Steve McQuerry 著．Cisco 现场手册：路由器配置 张辉译．北京：人民邮电出版社，2002．
[6] 张保通．网络互连技术——路由、交换与远程访问实训教程．北京：中国水利水电出版社，2010．
[7] 张保通，安志远．网络互连技术——路由、交换与远程访问．北京：中国水利水电出版社，2009．